ASM Ready Reference: Properties and Units for Engineering Alloys

Reviewed by the ASM International® Materials Properties Database Committee

J. Gilbert Kaufman, Chairman
The Aluminum Association

Bruce E. Boardman
Deere & Company

David R. Gerty
Charles Stark Draper Laboratory

Barry Hindin
Battelle Memorial Institute

Scott McCormick
ESM Software

Paul J. Sikorsky
The Trane Company

Gary A. Volk
Illinois Central College

First printing, August 1997

Great care is taken in the compilation and production of this Volume, but it should be made clear that NO WARRANTIES, EXPRESS OR IMPLIED, INCLUDING, WITHOUT LIMITATION, WARRANTIES OF MERCHANTABILITY OR FITNESS FOR A PARTICULAR PURPOSE, ARE GIVEN IN CONNECTION WITH THIS PUBLICATION. Although this information is believed to be accurate by ASM, ASM cannot guarantee that favorable results will be obtained from the use of this publication alone. This publication is intended for use by persons having technical skill, at their sole discretion and risk. Since the conditions of product or material use are outside of ASM's control, ASM assumes no liability or obligation in connection with any use of this information. No claim of any kind, whether as to products or information in this publication, and whether or not based on negligence, shall be greater in amount than the purchase price of this product or publication in respect of which damages are claimed. THE REMEDY HEREBY PROVIDED SHALL BE THE EXCLUSIVE AND SOLE REMEDY OF BUYER, AND IN NO EVENT SHALL EITHER PARTY BE LIABLE FOR SPECIAL, INDIRECT OR CONSEQUENTIAL DAMAGES WHETHER OR NOT CAUSED BY OR RESULTING FROM THE NEGLIGENCE OF SUCH PARTY. As with any material, evaluation of the material under end-use conditions prior to specification is essential. Therefore, specific testing under actual conditions is recommended.

Nothing contained in this book shall be construed as a grant of any right of manufacture, sale, use, or reproduction, in connection with any method, process, apparatus, product, composition, or system, whether or not covered by letters patent, copyright, or trademark, and nothing contained in this book shall be construed as a defense against any alleged infringement of letters patent, copyright, or trademark, or as a defense against liability for such infringement.

Comments, criticisms, and suggestions are invited, and should be forwarded to ASM International.

Library of Congress Cataloging-in-Publication Data:

ASM ready reference : properties and united for engineering alloys / reviewed by the ASM International Materials Properties Database Committee.
p. cm.
Includes bibliographical references and index.
ISBN 0-87170-585-0 (alk. paper)
1. Alloys. I. ASM International. Materials Properties Database Committee
TA48.A78 1997
620.1'6'021--dc21

97-13167
CIP

ISBN: 0-87170-585-0
SAN: 204-7586
ASM International®
Materials Park, OH 44073-0002
Printed in the United States of America

Introduction

This book was developed out of a need for consistency and equivalency when integrating material property data from multiple sources into reference print or electronic publications. The assembly of data bases in electronic format requires efficient decision making when assigning property values to a specific cell in a spread sheet or electronic table. The technical readers of books and journals may or may not be confused when comparing printed tables with column headings of F_{tu}, R_m, US, UTS, or Tenacity (and units of MPa, ksi or N/mm^2). Expectations when using a computer are more complex; one expects to find Tensile Strength by any other name, converted to the unit system of one's choice. For the computer to make life easier, someone must set up in the computer a thesaurus or lists of equivalent property names. Or, the person assembling a particular data base must know that, in that particular database, Tensile Strength is always "US" and always converted to MPa. This all sounds pretty easy to do, but many things contribute to the magnitude of the task. Some source material may include material and test parameters such as directionality in testing, test method, or quality indicators as part of a property name or abbreviation. Some properties, such as "Magnetic Permeability" or "Fracture Toughness" are inherently generic unless some details about test methodology are provided. This book is a beginner's guide, or a refresher course, for sorting apples to apples and helping to decide if Emittance is the same as Emissivity.

All readers of this book should be aware of a number of warnings. There is a temptation to paste stickers on the cover, similar to warning labels on medicine vials, such as:

* **Do not use unless you understand the concepts of significant digits and rounding.**
* **Stress, strain, magnetism, fracture mechanics, and any number of complex concepts are not defined or explained in this guide.**
* **This is not a standard or a specification.**
* **Use caution when substituting property names and abbreviations.**
* **It may not be possible to distinguish a property from a test method.**

There are a number of excellent references that anyone using material property data from multiple sources should consider. A few of these references are:

- E 380-93 Use of the International System of Units (SI) (the Modernized Metric System), American Society for Testing and Materials, West Conshohocken, PA, 1993
- NBS Special Publication 702, Standards and Metadata Requirements for Computerization of Selected Mechanical Properties of Metallic Materials, J.R. Westbrook, U.S. Department of Commerce, 1985
- Metals Handbooks, Volumes 1 and 2, ASM International, 1990, 1991

- Compilation of ASTM Standard Definitions, 8th Edition, American Society for Testing and Materials, West Conshohocken, PA, 1994
- ASM Materials Engineering Dictionary, J.R. Davis, ASM International, 1992
- SAE Handbook, Volume 1, Society of Automotive Engineers, 1997
- ASTM Standards on the Building of Materials Databases, sponsored by Committee E-49, American Society for Testing and Materials, 1993

The organization of this book is simple and straightforward. As indicated by the Table of Contents all information about a given material property is found on one page. ASTM standard definitions are in many cases incorporated into the definitions given for mechanical properties. The properties are arranged alphabetically within each Category. Then Index at the back of the book provides the page number for all property names, abbreviations, synonyms, symbols, and related terms.

Each property page outlines, when available, common usage for the following:

- Property name
- Common abbreviation
- Preferred unit
- Alternate unit
- Definition
- Example test method
- Material/Test parameters
- Conversion factors
- Synonyms, symbols, abbreviations, and related terms
- Example values

The Center for Materials Data at ASM International is grateful for review by the Materials Properties Databases Committee.

Contents

Bearing Strength

Common Abbreviation:	UBS
Preferred or SI Unit	MPa
Alternate or English Unit	ksi

DEFINITION:

Bearing Strength (UBS) is the maximum stress that sheet products can sustain when they are subjected to loading from external members joined to the sheet by riveting, bolting, or a similar fastening procedure. UBS is equal to the load at failure divided by the effective bearing area. Test results are reported for a specific ratio (e/D) of an edge distance (e) and a pin diameter (D). Typical e/D ratios are 1.5 or 2.0. Problems in testing with pin distortion, or failure, may be encountered when higher strength materials such as titanium and steel are tested. The results of bearing tests are influenced by the degree of lubrication on the pin. It is customary to use ultrasonic cleaning to prepare specimens and test fixtures, and to avoid touching the pin surface. F_{bru} is an abbreviation used for allowable Ultimate Bearing Stress.

EXAMPLE TEST METHOD: ASTM E 238

MATERIAL/TEST PARAMETERS: Edge distance, pin diameter, temperature, specimen width, specimen thickness, lubrication of pin surface

CONVERSION FACTORS:

to convert	to	multiply by	to convert	to	multiply by
MPa	ksi	0.14504	ksi	MPa	6.8948
psi	MPa	0.006895	MN/m^2	MPa	1
10^3 psi	MPa	6.895	N/mm^2	MPa	1
kgf/mm^2	MPa	9.807	tsi (short tons/in.^{2})	MPa	13.79
kgf/cm^2	MPa	980.7	tsi (long tons/in.^{2})	MPa	15.44

SYNONYMS, SYMBOLS, ABBREVIATIONS, AND RELATED TERMS: Ultimate Bearing Stress, F_{bru}, S_u

EXAMPLE VALUES:	**MPa**	**ksi**
Aluminum Alloys	300-1,000	40-130
Titanium Alloys	1,000-2,200	150-320

Bearing Yield Strength

Common Abbreviation:	BYS
Preferred or SI Unit	MPa
Alternate or English Unit	ksi

DEFINITION:

Bearing Yield Strength (BYS) is the bearing stress for the onset of nonlinear stress-strain behavior or permanent deformation. BYS depends on the e/D (edge distance/pin diameter) ratio of the bearing area, and on the specified limit for measuring deviations from the linear stress-strain relationship (for example, 0.2% offset). The results of bearing tests are influenced by the degree of lubrication on the pin. It is customary to use ultrasonic cleaning to prepare specimens and test fixtures, and to avoid touching the pin surface. F_{bry} is an abbreviation used for allowable Bearing Yield Stress.

EXAMPLE TEST METHOD: ASTM E 238

MATERIAL/TEST PARAMETERS: Offset, edge distance, pin diameter, lubrication of pin surface, temperature, specimen width, specimen thickness

CONVERSION FACTORS:

to convert	to	multiply by	to convert	to	multiply by
MPa	ksi	0.14504	ksi	MPa	6.8948
psi	MPa	0.006895	MN/m^2	MPa	1
10^3 psi	MPa	6.895	N/mm^2	MPa	1
kgf/mm^2	MPa	9.807	tsi (short tons/in.^{2})	MPa	13.79
kgf/cm^2	MPa	980.7	tsi (long tons/in.^{2})	MPa	15.44

SYNONYMS, SYMBOLS, ABBREVIATIONS, AND RELATED TERMS: Bearing Yield Stress, Bearing Proof Stress, Bearing Proof Strength, F_{bry}, S_y

EXAMPLE VALUES:	MPa	ksi
Aluminum Alloys	130-600	20-80
Titanium Alloys	350-1,700	50-250

Bending Strength

Common Abbreviation:	(None established)
Preferred or SI Unit	MPa
Alternate or English Unit	ksi

DEFINITION:

Bending Strength defines the bending moment or failure stress in bending of a material subjected to a bending load. Bending Strength is reported either as a bending moment (ASTM F 382) or as a calculated surface stress from a bending moment that produces permanent deformation in flat metallic products (ASTM E 855). Because of the relative complexity of stress-strain relationships in bending, Bending Strength usually is not quantified beyond that provided by the assumption of linear elastic behavior. F_b is an abbreviation used for allowable Bending Strength, or for the allowable Modulus of Rupture in Bending. The Modulus of Rupture in Bending is the value of maximum tensile or compressive stress (whichever causes failure) in the extreme fiber of a beam loaded to failure in bending.

EXAMPLE TEST METHODS: ASTM F 382, ASTM E 855

MATERIAL/TEST PARAMETERS: Test method, beam size, beam proportions

CONVERSION FACTORS:

to convert	to	multiply by	to convert	to	multiply by
MPa	ksi	0.14504	ksi	MPa	6.8948
psi	MPa	0.006895	MN/m^2	MPa	1
10^3 psi	MPa	6.895	N/mm^2	MPa	1
kgf/mm^2	MPa	9.807	tsi (short tons/in.^{2})	MPa	13.79
kgf/cm^2	MPa	980.7	tsi (long tons/in.^{2})	MPa	15.44

SYNONYMS, SYMBOLS, ABBREVIATIONS, AND RELATED TERMS: Bending Proof Strength, Offset Yield Strength in Bending, F_b, S_b, Modulus of Rupture in Bending

Compressive Strength

Common Abbreviation:	UCS
Preferred or SI Unit	MPa
Alternate or English Unit	ksi

DEFINITION:

Compressive Strength refers to the maximum compressive stress that a material can sustain without failure or unacceptable distortion. Brittle materials fail in compression by fracturing, and the compressive strength has a definite value. In the case of ductile, malleable, or semiviscous materials (which do not fail in compression by a shattering fracture), the value obtained for compressive strength is an arbitrary value dependent on the degree of distortion that is regarded as effective failure of the material. F_{cu} is an abbreviation used for allowable Ultimate Compressive Strength.

EXAMPLE TEST METHOD: ASTM E 9

MATERIAL/TEST PARAMETERS: Degree of distortion regarded as failure, specimen dimensions

CONVERSION FACTORS:

to convert	to	multiply by	to convert	to	multiply by
MPa	ksi	0.14504	ksi	MPa	6.8948
psi	MPa	0.006895	MN/m^2	MPa	1
10^3 psi	MPa	6.895	N/mm^2	MPa	1
kgf/mm^2	MPa	9.807	tsi (short tons/in.^{2})	MPa	13.79
kgf/cm^2	MPa	980.7	tsi (long tons/in.^{2})	MPa	15.44

SYNONYMS, SYMBOLS, ABBREVIATIONS, AND RELATED TERMS: Ultimate Compressive Stress, S_{uc}, R_{cm}, F_{cu}

Compressive Yield Strength

Common Abbreviation:	CYS
Preferred or SI Unit	MPa
Alternate or English Unit	ksi

DEFINITION:

Compressive Yield Strength is the compressive stress for the onset of nonlinear stress-strain behavior, usually measured at an offset of 0.2% on the compressive stress-strain curve. The measured yield strength depends on the specified strain limit for measuring permanent deformation (offset) or the deviation from a linear stress-strain relationship. Compressive Yield Strengths are affected by previous tensile strains (the Bauschinger effect). F_{cy} is an abbreviation used for allowable Compressive Yield Stress at which permanent strain equals 0.002.

EXAMPLE TEST METHOD: ASTM E 9

MATERIAL/TEST PARAMETERS: Specified strain limit (offset)

CONVERSION FACTORS:

to convert	to	multiply by	to convert	to	multiply by
MPa	ksi	0.14504	ksi	MPa	6.8948
psi	MPa	0.006895	MN/m^2	MPa	1
10^3 psi	MPa	6.895	N/mm^2	MPa	1
kgf/mm^2	MPa	9.807	tsi (short tons/in.^{2})	MPa	13.79
kgf/cm^2	MPa	980.7	tsi (long tons/in.^{2})	MPa	15.44

SYNONYMS, SYMBOLS, ABBREVIATIONS, AND RELATED TERMS:
Compressive Proof Stress, S_{cy}, Compressive Proof Strength, R_{cy}, $R_{c0.2}$, F_{cy}, Compressive Yield Stress

EXAMPLE VALUES:	MPa	ksi
Aluminum Alloys	70-560	10-80

Activation Energy for Creep

Common Abbreviation:	Q_c
Preferred or SI Unit	kJ/mole
Alternate or English Unit	kcal/mole

DEFINITION:

Activation Energy for Creep (Q_c) is a parameter in an Arrhenius-type model of the temperature dependence for steady-state creep rates. Above approximately 1.6 T_m (where T_m is the absolute melting point of the metal or alloy), Q_c is independent of temperature and is equal to the activation energy for self-diffusion (Q_D). Measured activation energies vary significantly with temperature at low temperatures, and become independent of temperature at approximately 0.6 T_m. Low activation energies indicate a weak temperature dependence of creep.

MATERIAL/TEST PARAMETERS: Temperature

CONVERSION FACTORS:

to convert	to	multiply by	to convert	to	multiply by
kJ/mole	kcal/mole	0.239	kcal/mole	kJ/mole	4.184

SYNONYMS, SYMBOLS, ABBREVIATIONS, AND RELATED TERMS: ΔH

Creep Rate

Common Abbreviation: $\dot{\varepsilon}$

Preferred or SI Unit (mm/mm)/h
Alternate or English Unit (in./in.)/h

DEFINITION:

Creep Rate is the rate at which a material deforms over time due to an externally applied constant load. Creep Rate is the slope at any given time on a creep-time curve (Creep strain versus Time). Average creep rate is the average slope.

EXAMPLE TEST METHOD: ASTM E 139

MATERIAL/TEST PARAMETERS: Load, test method, time, temperature, gage length

CONVERSION FACTORS:

to convert	to	multiply by	to convert	to	multiply by
(mm/mm)/h	(in./in.)/h	1	(in./in.)/h	(mm/mm)/h	1

SYNONYMS, SYMBOLS, ABBREVIATIONS, AND RELATED TERMS: dε/dt, Rate of Creep, V_k

Creep Strain

Common Abbreviation:	ε_c
Preferred or SI Unit	mm/mm
Alternate or English Unit	in./in.

DEFINITION:

Creep refers to the continuous (time-dependent) deformation of materials due to thermal activation and the application of stress. Creep Strain is often graphed as a function of time, and the curves of creep-time plots are divided into regions of Primary Creep, Secondary Creep, and Tertiary Creep. Creep tests are usually made at constant load and temperature. The initial strain (before creep occurs) is reported in plastic tests but not in metal tests.

EXAMPLE TEST METHOD: ASTM E 139

MATERIAL/TEST PARAMETERS: Load, temperature, gage length, specified time during testing

CONVERSION FACTORS:

to convert	to	multiply by	to convert	to	multiply by
mm/mm	in./in.	1	in./in.	mm/mm	1

SYNONYMS, SYMBOLS, ABBREVIATIONS, AND RELATED TERMS: Creep

Creep Strength

Common Abbreviation:	(None established)
Preferred or SI Unit	MPa
Alternate or English Unit	ksi

DEFINITION:

Creep Strength is the stress causing a given creep strain in a creep test, at any given time in a specified constant temperature and environment. Creep Strength is commonly reported for strains of 0.1, 0.2, 0.5, 1.0, 2.0, and 5.0%, and at times of 1,000, 10,000, and 100,000 .

MATERIAL/TEST PARAMETERS: Temperature, strain, time

CONVERSION FACTORS:

to convert	to	multiply by	to convert	to	multiply by
MPa	ksi	0.14504	ksi	MPa	6.8948
psi	MPa	0.006895	MN/m^2	MPa	1
10^3 psi	MPa	6.895	N/mm^2	MPa	1
kgf/mm^2	MPa	9.807	tsi (short tons/in.^{2})	MPa	13.79
kgf/cm^2	MPa	980.7	tsi (long tons/in.^{2})	MPa	15.44

Creep-Rupture Ductility (%El)

Common Abbreviation:	(None established)
Preferred or SI Unit	%El
Alternate or English Unit	%El

DEFINITION:

Creep-Rupture Ductility is a measure of a material's ability to deform prior to rupture in a creep-rupture test. Units of percent elongation (%El) for the degree of extension at rupture, or units of percent reduction of area (%RA) may be used, or, both may be reported.

EXAMPLE TEST METHOD: ASTM E 139

MATERIAL/TEST PARAMETERS: Time, temperature, gage length, strain rate, test method

SYNONYMS, SYMBOLS, ABBREVIATIONS, AND RELATED TERMS: Rupture Ductility, Creep Ductility

Creep-Rupture Ductility (%RA)

Common Abbreviation:	(None established)
Preferred or SI Unit	%RA
Alternate or English Unit	%RA

DEFINITION:

Creep-Rupture Ductility is a measure of a material's ability to deform prior to rupture in a creep-rupture test. Units of percent elongation (%El) for the degree of extension at rupture, or units of percent reduction of area (%RA) may be used, or, both may be reported.

EXAMPLE TEST METHOD: ASTM E 139

MATERIAL/TEST PARAMETERS: Time, temperature, strain rate, test method

SYNONYMS, SYMBOLS, ABBREVIATIONS, AND RELATED TERMS: Rupture Ductility, Creep Ductility

Creep-Rupture Strength

Common Abbreviation:	σ_r
Preferred or SI Unit	MPa
Alternate or English Unit	ksi

DEFINITION:

Creep-Rupture Strength is the stress that will cause fracture or rupture in a creep test. Creep-Rupture Strength is typically used to measure a material's elevated temperature durability.

EXAMPLE TEST METHOD: ASTM E 139

MATERIAL/TEST PARAMETERS: Time (usually reported at 100, 1,000, 10,000, and 100,000 h), temperature

CONVERSION FACTORS:

to convert	to	multiply by	to convert	to	multiply by
MPa	ksi	0.14504	ksi	MPa	6.8948
psi	MPa	0.006895	MN/m^2	MPa	1
10^3 psi	MPa	6.895	N/mm^2	MPa	1
kgf/mm^2	MPa	9.807	tsi (short tons/$in.^2$)	MPa	13.79
kgf/cm^2	MPa	980.7	tsi (long tons/$in.^2$)	MPa	15.44

SYNONYMS, SYMBOLS, ABBREVIATIONS, AND RELATED TERMS: Rupture Stress, Creep-Rupture Stress, Breaking Stress, Rupture Strength, Stress-Rupture Strength

Notch Rupture Strength

Common Abbreviation:	(None established)
Preferred or SI Unit	MPa
Alternate or English Unit	ksi

DEFINITION:

Notch Rupture Strength is the stress that will cause fracture or rupture in a creep test of a notched specimen. Notch Rupture Strength is calculated as the ratio of the applied load to the original area of the minimum cross section in a stress-rupture test of a notched specimen.

EXAMPLE TEST METHOD: ASTM E 292

MATERIAL/TEST PARAMETERS: Time, temperature, load, notch geometry

CONVERSION FACTORS:

to convert	to	multiply by	to convert	to	multiply by
MPa	ksi	0.14504	ksi	MPa	6.8948
psi	MPa	0.006895	MN/m^2	MPa	1
10^3 psi	MPa	6.895	N/mm^2	MPa	1
kgf/mm^2	MPa	9.807	tsi (short tons/in.^{2})	MPa	13.79
kgf/cm^2	MPa	980.7	tsi (long tons/in.^{2})	MPa	15.44

SYNONYMS, SYMBOLS, ABBREVIATIONS, AND RELATED TERMS: Notch Stress Rupture Strength

Primary Creep

Common Abbreviation:	(None established)
Preferred or SI Unit	mm/mm
Alternate or English Unit	in./in.

DEFINITION:

Primary Creep is the strain occurring upon initial loading in a creep test. Primary Creep occurs during a period of decreasing creep rate until the steady-state creep rate of Secondary Creep is reached. The decreasing Creep Rate during Primary Creep is attributed to strain hardening.

EXAMPLE TEST METHOD: ASTM E 139

MATERIAL/TEST PARAMETERS: Time, temperature, load

CONVERSION FACTORS:

to convert	to	multiply by	to convert	to	multiply by
mm/mm	in./in.	1	in./in.	mm/mm	1

SYNONYMS, SYMBOLS, ABBREVIATIONS, AND RELATED TERMS: First-Stage Creep

Rupture Life

Common Abbreviation:	t_r
Preferred or SI Unit	h
Alternate or English Unit	h

DEFINITION:

Rupture Life is the time to rupture obtained from creep-rupture testing at a constant stress and temperature.

MATERIAL/TEST PARAMETERS: Initial stress, initial extension, creep extension, temperature

SYNONYMS, SYMBOLS, ABBREVIATIONS, AND RELATED TERMS: Time to Rupture

Secondary Creep

Common Abbreviation:	ε_s
Preferred or SI Unit	mm/mm
Alternate or English Unit	in./in.

DEFINITION:

Secondary Creep is the strain occurring at a nearly constant creep rate after the primary creep stage has reached its minimum creep rate. Secondary Creep is explained in terms of a balance between strain hardening and the softening and damage processes, resulting in a nearly constant creep rate. Minimum Creep Rate and Steady-State Creep are often synonymous with Secondary Creep because creep rates are at a minimum and nearly constant during Secondary Creep.

EXAMPLE TEST METHOD: ASTM E 139

MATERIAL/TEST PARAMETERS: Time, temperature, load

CONVERSION FACTORS:

to convert	to	multiply by	to convert	to	multiply by
mm/mm	in./in.	1	in./in.	mm/mm	1

SYNONYMS, SYMBOLS, ABBREVIATIONS, AND RELATED TERMS: Second-Stage Creep, Steady-State Creep, Minimum Creep Rate

Tertiary Creep

Common Abbreviation:	(None established)
Preferred or SI Unit	mm/mm
Alternate or English Unit	in./in.

DEFINITION:

Tertiary Creep is the third stage of creep behavior that follows the Secondary or Minimum Creep Rate, and increases with time, up to rupture. The acceleration of creep rates is associated with the increase of damage mechanisms, and tertiary creep often is distinguished from secondary creep by the onset of necking.

CONVERSION FACTORS:

to convert	to	multiply by	to convert	to	multiply by
mm/mm	in./in.	1	in./in.	mm/mm	1

SYNONYMS, SYMBOLS, ABBREVIATIONS, AND RELATED TERMS: Third-Stage Creep

Acoustic Dissipation

Common Abbreviation: Ω

Preferred or SI Unit (Unitless)

DEFINITION:

Acoustic Dissipation (Ω) is defined by the relation: $\Omega = P_a - P_{tr}$, where P_a is the absorbed sound energy flux and P_{tr} is the transmitted component. Acoustic dissipation is more pronounced at high frequencies. For most materials, dissipation increases directly with frequency.

MATERIAL/TEST PARAMETERS: Temperature, frequency

SYNONYMS, SYMBOLS, ABBREVIATIONS, AND RELATED TERMS:
Dissipation Factor (acoustic), Ultrasonic Dissipation, Attenuation (acoustic)

Damping Capacity

Common Abbreviation:	Q^{-1}
Preferred or SI Unit	Logarithmic Decrement
Alternate or English Unit	dB/cycle

DEFINITION:

Damping Capacity is the ability of a material to absorb vibrational energy by internal friction, converting the mechanical energy into heat. Damping Capacity can be expressed in units of logarithmic decrement of applied stress, which is defined as the natural logarithm (ln) of the ratio of the amplitude of one oscillation to that after one period of oscillation. Internal Friction generally refers to low strain levels of vibration, while Damping Capacity has a broader meaning that includes stresses near the yield strength.

MATERIAL/TEST PARAMETERS: Stress amplitude, frequency of vibration, temperature

CONVERSION FACTORS:

Logarithmic Decrement = $\ln[10^{dB}]$

dB/cycle = log[exp(Log Decrement)]

SYNONYMS, SYMBOLS, ABBREVIATIONS, AND RELATED TERMS:
Attenuation, Internal Friction

Damping Factor

Common Abbreviation:	(None established)
Preferred or SI Unit	s^{-1}
Alternate or English Unit	s^{-1}

DEFINITION:

The Damping Factor is the logarithmic decrement of any underdamped harmonic motion divided by the time period of one oscillation. Logarithmic decrement is defined as the natural logarithm (ln) of the ratio of the amplitude of one oscillation to that after one period of oscillation.

MATERIAL/TEST PARAMETERS: Temperature

SYNONYMS, SYMBOLS, ABBREVIATIONS, AND RELATED TERMS: Damping Coefficient, Damping Resistance

Flow Stress

Common Abbreviation:	Ω
Preferred or SI Unit	MPa
Alternate or English Unit	ksi

DEFINITION:

Flow Stress is the true stress needed to produce plastic deformation.

MATERIAL/TEST PARAMETERS: Temperature, strain rate

CONVERSION FACTORS:

to convert	to	multiply by	to convert	to	multiply by
MPa	ksi	0.14504	ksi	MPa	6.8948
psi	MPa	0.006895	MN/m^2	MPa	1
10^3 psi	MPa	6.895	N/mm^2	MPa	1
kgf/mm^2	MPa	9.807	tsi (short tons/in.^{2})	MPa	13.79
kgf/cm^2	MPa	980.7	tsi (long tons/in.^{2})	MPa	15.44

Hall-Petch Friction Stress

Common Abbreviation:	Ω_i
Preferred or SI Unit	MPa
Alternate or English Unit	ksi

DEFINITION:

Hall-Petch Friction Stress is a parameter used to model the effect of grain size on yield stress, flow stress, and fracture stress. See also Hall-Petch Grain Size Constant and Fracture Grain Size Constant.

CONVERSION FACTORS:

to convert	to	multiply by	to convert	to	multiply by
MPa	ksi	0.14504	ksi	MPa	6.8948
psi	MPa	0.006895	MN/m^2	MPa	1
10^3 psi	MPa	6.895	N/mm^2	MPa	1
kgf/mm^2	MPa	9.807	tsi (short tons/in.^{2})	MPa	13.79
kgf/cm^2	MPa	980.7	tsi (long tons/in.^{2})	MPa	15.44

SYNONYMS, SYMBOLS, ABBREVIATIONS, AND RELATED TERMS: Friction Stress

Hall-Petch Grain Size Constant

Common Abbreviation:	K
Preferred or SI Unit	MPa / $\sqrt{\text{mm}}$
Alternate or English Unit	ksi / $\sqrt{\text{in.}}$

DEFINITION:

The Hall-Petch Grain Size Constant is a parameter, K, used to model the effect of grain size on both yield stress and flow stress (Ω) such that $\Omega = \Omega_i + Kd^{(1/2)}$, where Ω_i is the Hall-Petch Friction Stress, and d is the average grain diameter.

CONVERSION FACTORS:

to convert	to	multiply by
MPa / $\sqrt{\text{mm}}$	ksi / $\sqrt{\text{in.}}$	0.07309
ksi / $\sqrt{\text{in.}}$	MPa / $\sqrt{\text{mm}}$	1.368
psi / $\sqrt{\text{in.}}$	MPa / $\sqrt{\text{mm}}$	0.001368
kgf / $\sqrt{\text{mm}}$	MPa / $\sqrt{\text{mm}}$	9.807

Strain-Hardening Exponent

Common Abbreviation: n

Preferred or SI Unit (Unitless)

DEFINITION:

The Strain-Hardening Exponent is the value of n in the relationship $\sigma = K\varepsilon^n$, where σ is the true stress, ε is the true strain, and K is equal to the true stress at a true strain of 1.0. The Strain-Hardening Exponent is equal to the slope of the true stress/true strain curve up to a maximum load, when plotted on log-log coordinates. The property relates to the ability of a material to be stretched in a metalworking operation. In general, the higher the value, the better a material's formability. For example, formable sheet steels have Strain-Hardening Exponents greater than 0.18. Values as low as 0.10 are typical of hot rolled and other steels not processed for formability.

EXAMPLE TEST METHODS: ASTM E 646, SAE J877

MATERIAL/TEST PARAMETERS: Temperature

SYNONYMS, SYMBOLS, ABBREVIATIONS, AND RELATED TERMS: Stretch Formability, n-value, Monotonic Strain-Hardening Exponent, Work-Hardening Factor

EXAMPLE VALUES: **(Unitless)**

Carbon and Alloy Steels	0.10-0.48
Aluminum Alloys	0.1-0.3
Copper Alloys	0.49-0.56

Strain-Rate Sensitivity

Common Abbreviation:	m
Preferred or SI Unit	(Unitless)

DEFINITION:

The Strain-Rate Exponent is the slope of a log-log plot of flow stress versus the strain rate. The Strain-Rate Exponent is also called the Strain-Rate Sensitivity of the flow stress, and reflects the fact that an increase in strain rate increases flow stress (strain-rate hardening). For most materials, the effect is modest for cold working, but is quite significant for hot working.

EXAMPLE TEST METHOD: SAE J877

MATERIAL/TEST PARAMETERS: Temperature, strain level

SYNONYMS, SYMBOLS, ABBREVIATIONS, AND RELATED TERMS: Strain-Rate Exponent, Strain-Rate Sensitivity Exponent, Strain-Rate Hardening, m-value

EXAMPLE VALUES: (Unitless)

Carbon and Alloy Steels	0.007 to 0.015
Aluminum Alloys	–0.002 to 0.005
Copper Alloys	0.001

Strength Coefficient

Common Abbreviation:	K
Preferred or SI Unit	MPa
Alternate or English Unit	ksi

DEFINITION:

The Strength Coefficient is equal to the true stress at a true strain of 1.0.

MATERIAL/TEST PARAMETERS: Temperature

CONVERSION FACTORS:

to convert	to	multiply by	to convert	to	multiply by
MPa	ksi	0.14504	ksi	MPa	6.8948
psi	MPa	0.006895	MN/m^2	MPa	1
10^3 psi	MPa	6.895	N/mm^2	MPa	1
kgf/mm^2	MPa	9.807	tsi (short tons/in.^{2})	MPa	13.79
kgf/cm^2	MPa	980.7	tsi (long tons/in.^{2})	MPa	15.44

SYNONYMS, SYMBOLS, ABBREVIATIONS, AND RELATED TERMS:
Monotonic Strength Coefficient

EXAMPLE VALUES:	**MPa**	**ksi**
Carbon and Alloy Steels	525-1575	75-225
Copper Alloys	320-900	45-130

Bulk Modulus

Common Abbreviation:	K
Preferred or SI Unit	GPa
Alternate or English Unit	10^6 psi

DEFINITION:

Bulk Modulus (K) is a measure of resistance to volume change. K is expressed as the ratio of hydrostatic pressure to the corresponding unit change in volume. The Bulk Modulus is the reciprocal of Compressibility when stresses do not exceed the proportional limit. K is related to Young's Modulus (E), and to Poisson's Ratio (ν) by the equation: $E = 3K(1 - 2\nu)$.

MATERIAL/TEST PARAMETERS: Temperature

CONVERSION FACTORS:

to convert	to	multiply by	to convert	to	multiply by
GPa	10^6 psi	0.145	10^6 psi	GPa	6.895
MPa	GPa	0.001	MN/m^2	GPa	1,000
ksi	GPa	0.006895	N/mm^2	GPa	1,000
10^3 ksi	GPa	6.895	tsi(short tons/in.^{2})	GPa	0.01379
kgf/cm^2	GPa	0.0000981	kgf/mm^2	GPa	0.00981

SYNONYMS, SYMBOLS, ABBREVIATIONS, AND RELATED TERMS:
Compression Modulus, Modulus of Compression, Hydrostatic Modulus, Volumetric Modulus of Elasticity, Bulk Modulus of Elasticity, Modulus in Compression

EXAMPLE VALUES:	**GPa**	**10^6 psi**
Carbon and Alloy Steels	165.6-169	24-24.5
Aluminum Alloys	75.2	10.9
Copper Alloys	111.8-138	16.2-20
Magnesium Alloys	35.6	5.16
Stainless Steels	166.3	24.1
Titanium Alloys	108.3	15.7

Chord Modulus

Common Abbreviation:	(None established)
Preferred or SI Unit	GPa
Alternate or English Unit	10^6 psi

DEFINITION:

The Chord Modulus is the slope of a chord drawn between any two specified points below the elastic limit on a stress-strain curve. The property is used instead of Young's Modulus for materials that do not conform to Hooke's Law. (A material in which the stress is linearly proportional to strain in a stress-strain curve is said to conform to Hooke's Law.)

EXAMPLE TEST METHOD: ASTM E 111

MATERIAL/TEST PARAMETERS: Temperature

CONVERSION FACTORS:

to convert	to	multiply by	to convert	to	multiply by
GPa	10^6 psi	0.145	10^6 psi	Gpa	6.895
MPa	GPa	0.001	MN/m^2	GPa	1,000
ksi	GPa	0.006895	N/mm^2	GPa	1,000
10^3 ksi	GPa	6.895	tsi	GPa	0.01379
kgf/cm^2	GPa	0.0000981	kgf/mm^2	GPa	0.00981

Poisson's Ratio

Common Abbreviation:	ν
Preferred or SI Unit	(Unitless)

DEFINITION:

Poisson's Ratio (ν) is the absolute value of the ratio of transverse (lateral) strain to the corresponding axial strain resulting from uniformly distributed axial stress below the proportional limit of the material. Poisson's Ratio is related to Young's Modulus (E) and the Shear Modulus (G) by the relation: $\nu = (E/2G) - 1$. Most metals have Poisson's Ratios of 0.25 to 0.40. The ratio depends on crystallographic orientation for nonisotropic materials. Above the proportional limit, the ratio of transverse strain to axial strain will depend on the average stress and stress range for which it is measured, and should not be regarded as Poisson's Ratio.

EXAMPLE TEST METHOD: ASTM E 132

MATERIAL/TEST PARAMETERS: Temperature

SYNONYMS, SYMBOLS, ABBREVIATIONS, AND RELATED TERMS: Poisson Ratio

EXAMPLE VALUES:	**(Unitless)**
Carbon and Alloy Steels	0.29-0.3
Aluminum Alloys	0.33
Copper Alloys	0.34-0.35
Magnesium Alloys	0.29
Stainless Steels	0.27-0.3
Titanium Alloys	0.26-0.36

Secant Modulus

Common Abbreviation:	(None established)
Preferred or SI Unit	GPa
Alternate or English Unit	10^6 psi

DEFINITION:

The Secant Modulus is the slope of a secant from the origin to any specified point on the stress-strain curve. The property can be used instead of Young's Modulus for materials that do not conform to Hooke's Law. (A material in which the stress is linearly proportional to strain in a stress-strain curve is said to obey Hooke's Law.)

MATERIAL/TEST PARAMETERS: Temperature

CONVERSION FACTORS:

to convert	to	multiply by	to convert	to	multiply by
GPa	10^6 psi	0.145	10^6 psi	GPa	6.895
MPa	GPa	0.001	MN/m^2	GPa	1000
ksi	GPa	0.006895	N/mm^2	GPa	1000
10^3 ksi	GPa	6.895	tsi	GPa	0.01379
kgf/cm^2	GPa	0.0000981	kgf/mm^2	GPa	0.00981

Shear Modulus

Common Abbreviation:	G
Preferred or SI Unit	GPa
Alternate or English Unit	10^6 psi

DEFINITION:

The Shear Modulus (G) is the ratio of shear stress to the corresponding shear strain for elastic shear stresses below the proportional limit. G is usually determined by torsion testing. For isotropic materials, G is related to Young's Modulus (E), and to Poisson's Ratio (ν) by the equation $G = E/2(1 + \nu)$.

EXAMPLE TEST METHOD: ASTM E 143

MATERIAL/TEST PARAMETERS: Temperature, stress range, direction if material is anisotropic

CONVERSION FACTORS:

to convert	to	multiply by	to convert	to	multiply by
GPa	10^6 psi	0.145	10^6 psi	GPa	6.895
MPa	GPa	0.001	MN/m^2	GPa	1000
ksi	GPa	0.006895	N/mm^2	GPa	1000
10^3 ksi	GPa	6.895	tsi	GPa	0.01379
kgf/cm^2	GPa	0.0000981	kgf/mm^2	GPa	0.00981

SYNONYMS, SYMBOLS, ABBREVIATIONS, AND RELATED TERMS: Torsional Modulus, Modulus of Rigidity, Shear Modulus of Elasticity

EXAMPLE VALUES:	**GPa**	**10^6 psi**
Carbon and Alloy Steels	78-82.1	11.3-11.9
Aluminum Alloys	26.2	3.8
Copper Alloys	37.3-48.4	5.41-7.01
Magnesium Alloys	17.3	2.51
Stainless Steels	84.2	12.2
Titanium Alloys	45.6	6.61

Tangent Modulus

Common Abbreviation:	(None established)
Preferred or SI Unit	GPa
Alternate or English Unit	10^6 psi

DEFINITION:

The Tangent Modulus is the slope of the stress-strain curve at a specified value of stress or strain. The property can be used instead of Young's Modulus for materials that do not conform to Hooke's Law. (A material in which stress is linearly proportional to strain is said to obey Hooke's Law.)

EXAMPLE TEST METHOD: ASTM E 111

MATERIAL/TEST PARAMETERS: Temperature, stress or strain value

CONVERSION FACTORS:

to convert	to	multiply by	to convert	to	multiply by
GPa	10^6 psi	0.145	10^6 psi	GPa	6.895
MPa	GPa	0.001	MN/m^2	GPa	1,000
ksi	GPa	0.006895	N/mm^2	GPa	1,000
10^3 ksi	GPa	6.895	tsi	GPa	0.01379
kgf/cm^2	GPa	0.0000981	kgf/mm^2	GPa	0.00981

Young's Modulus

Common Abbreviation:	E
Preferred or SI Unit	GPa
Alternate or English Unit	10^6 psi

DEFINITION:

Young's Modulus (E, or the Modulus of Elasticity) is a measure of rigidity or stiffness of a material, and is defined as the ratio of stress to strain in the elastic region. E is numerically equal to the slope of the stress-strain curve in the range of linear proportionality of stress to strain. (A material in which the stress is linearly proportional to strain is said to obey Hooke's Law.) Young's Modulus is a term sometimes used for both compressive and tensile stresses, provided that the strain is directly proportional to the applied stress. Nonisotropic materials may possess greatly different compressive and tensile stress-strain curves, and, therefore, different bulk and elastic moduli.

EXAMPLE TEST METHOD: ASTM E 111

MATERIAL/TEST PARAMETERS: Temperature, crystallographic orientation (texture)

CONVERSION FACTORS:

to convert	to	multiply by	to convert	to	multiply by
GPa	10^6 psi	0.145	10^6 psi	GPa	6.895
MPa	GPa	0.001	MN/m^2	GPa	1,000
ksi	GPa	0.006895	N/mm^2	GPa	1,000
10^3 ksi	GPa	6.895	tsi	GPa	0.01379
kgf/cm^2	GPa	0.0000981	kgf/mm^2	GPa	0.00981

SYNONYMS, SYMBOLS, ABBREVIATIONS, AND RELATED TERMS: Elastic Modulus, Elastic Tensile Modulus, Modulus of Elasticity, Coefficient of Elasticity, Static Modulus, Tensile Modulus, Modulus in Tension

EXAMPLE VALUES:	**GPa**	**10^6 psi**
Carbon and Alloy Steels	200-212	29-31
Aluminum Alloys	70	10
Copper Alloys	100-133	15-19
Magnesium Alloys	45	6.5
Stainless Steels	190-215	27.5-31
Titanium Alloys	120	17.5

Cyclic Strain-Hardening Exponent

Common Abbreviation: n'

Preferred or SI Unit (Unitless)

DEFINITION:

The Cyclic Strain-Hardening Exponent (n') is the power to which true plastic strain amplitude must be raised in order to be proportional to true stress amplitude, according to the equation $\Delta\sigma/2 = K'(\Delta\varepsilon_p/2)^{n'}$ where $(\Delta\varepsilon_p/2)$ is the true plastic strain amplitude, K' is the Cyclic Strength Coefficient, and $\Delta\sigma$ is the stress range. In general, metals with exponents >0.15 cyclically harden; those with exponents <0.15 cyclically soften. Cyclic softening is typical of many cold-worked pure metals and steels at small strain amplitudes. Annealed pure metals, many aluminum alloys, and as-quenched steels cyclically harden.

EXAMPLE TEST METHOD: SAE J1099

Cyclic Strength Coefficient

Common Abbreviation:	K′
Preferred or SI Unit	MPa
Alternate or English Unit	ksi

DEFINITION:

The Cyclic Strength Coefficient is the true stress at a true plastic strain of unity in the equation $\Delta\sigma/2 = K'(\Delta\varepsilon_p/2)^{n'}$ where $(\Delta\varepsilon_p/2)$ is the true plastic strain amplitude, n′ is the Cyclic Strain-Hardening Exponent, and $\Delta\sigma$ is the stress range.

EXAMPLE TEST METHOD: SAE J1099

CONVERSION FACTORS:

to convert	to	multiply by	to convert	to	multiply by
MPa	ksi	0.14504	ksi	MPa	6.8948
psi	MPa	0.006895	MN/m^2	MPa	1
10^3 psi	MPa	6.895	N/mm^2	MPa	1
kgf/mm^2	MPa	9.807	tsi (short tons/in.^{2})	MPa	13.79
kgf/cm^2	MPa	980.7	tsi (long tons/in.^{2})	MPa	15.44

Cyclic Yield Strength

Common Abbreviation:	σ_{ys}
Preferred or SI Unit	MPa
Alternate or English Unit	ksi

DEFINITION:

Cyclic Yield Strength is Yield Strength measured during cyclic loading. In many cases Cyclic Yield Strength is determined at the same 0.2% offset as the static or monotonic Tensile Yield Strength. If a metal softens during cyclic loading, then the Cyclic Yield Strength will be lower than the static (monotonic) Yield Strength. Cyclic softening and a reduction in yield strength during cyclic loading is typical of many cold-worked pure metals and steels at small strain amplitudes. Annealed pure metals, many aluminum alloys, and as-quenched steels cyclically harden.

EXAMPLE TEST METHOD: SAE J1099

MATERIAL/TEST PARAMETERS: Offset, temperature

CONVERSION FACTORS:

to convert	to	multiply by	to convert	to	multiply by
MPa	ksi	0.14504	ksi	MPa	6.8948
psi	MPa	0.006895	MN/m^2	MPa	1
10^3 psi	MPa	6.895	N/mm^2	MPa	1
kgf/mm^2	MPa	9.807	tsi (short tons/in.^{2})	MPa	13.79
kgf/cm^2	MPa	980.7	tsi (long tons/in.^{2})	MPa	15.44

SYNONYMS, SYMBOLS, ABBREVIATIONS, AND RELATED TERMS: Yield Strength in Fatigue

Fatigue Crack Growth Rate

Common Abbreviation:	da/dN
Preferred or SI Unit	mm/cycle
Alternate or English Unit	in./cycle

DEFINITION:

The Fatigue Crack Growth Rate is the rate of crack extension caused by constant-amplitude fatigue loading, expressed in terms of crack extension per cycle for a specified stress-intensity range (ΔK). In many cases of crack growth rate testing, the Fatigue Crack Growth Rate (da/dN) may be related to the stress-intensity factor by the relationship $da/dN = C(\Delta K)^n$ where ΔK is the stress-intensity factor range, and C and n are constants for a given material and stress ratio. C and n can be obtained from the intercept and slope, respectively, of the linear log da/dN versus logΔK plot. Measured values of da/dN are helpful for determining the conditions for subcritical crack growth, or the crack size-load combinations that will cause cracks to grow from small initial sizes to critical sizes.

EXAMPLE TEST METHOD: ASTM E 647

MATERIAL/TEST PARAMETERS: Stress ratio, temperature, stress-intensity factor, cycling rate

CONVERSION FACTORS:

to convert	to	multiply by
cm/cycle	mm/cycle	10

SYNONYMS, SYMBOLS, ABBREVIATIONS, AND RELATED TERMS: Fatigue Crack Propagation Rate, Crack Growth per Cycle, Crack Growth Rate, $\Delta a/\Delta N$

Fatigue Ductility Coefficient

Common Abbreviation: ε'_f

Preferred or SI Unit (Unitless)

DEFINITION:

The Fatigue Ductility Coefficient (ε'_f) is the true strain required to cause failure in one reversal. ε'_f is the intercept, at $2N_f = 1$, on a log-log plot of $\Delta\varepsilon_p/2$ (plastic strain amplitude) versus $2N_f$ (reversals to failure, or twice the number of cycles to failure). The Fatigue Ductility Coefficient is used with the Fatigue Ductility Exponent to determine the elastic portion of the Fatigue Life.

EXAMPLE TEST METHOD: SAE J1099

Fatigue Ductility Exponent

Common Abbreviation: c

Preferred or SI Unit (Unitless)

DEFINITION:

The Fatigue Ductility Exponent (c) is the power to which Fatigue Life (expressed in reversals) must be raised in order to be proportional to the true strain amplitude. The Fatigue Ductility Exponent is the slope of a log-log plot of $\Delta\varepsilon_p/2$ (plastic strain amplitude) versus $2N_f$ (reversals to failure, or twice the number of cycles to failure). The Fatigue Ductility Exponent is used with the Fatigue Ductility Coefficient to determine the elastic portion of Fatigue Life.

EXAMPLE TEST METHOD: SAE J1099

Fatigue Endurance Ratio

Common Abbreviation: (None established)

Preferred or SI Unit (Unitless)

DEFINITION:

The Fatigue Endurance Ratio is the ratio of Fatigue Limit and Tensile Strength.

MATERIAL/TEST PARAMETERS: Specimen type (notched or unnotched)

Fatigue Life (Cycles)

Common Abbreviation:	N
Preferred or SI Unit	Number of Cycles
Alternate or English Unit	Number of Reversals

DEFINITION:

Fatigue life expresses the ability of a material to withstand a given cyclic exposure. Fatigue Life (Cycles) is a measure of the cycles or reversals to failure for a specimen due to the effects of fatigue at a given stress or strain amplitude.

MATERIAL/TEST PARAMETERS: Stress or strain amplitude, mean stress or strain, cycling frequency, wave form of the stress or strain cycle, temperature

CONVERSION FACTORS:

to convert	to	multiply by
Number of Cycles	Number of Reversals	2
Number of Reversals	Number of Cycles	0.5

SYNONYMS, SYMBOLS, ABBREVIATIONS, AND RELATED TERMS: Cycles to Failure, Reversals to Failure

Fatigue Life (Time)

Common Abbreviation:	(None established)
Preferred or SI Unit	h
Alternate or English Unit	h

DEFINITION:

Fatigue Life (Time) is a measure of the time to failure for a specimen due to the effects of fatigue at a given stress or strain amplitude.

MATERIAL/TEST PARAMETERS: Stress or strain amplitude, mean stress or strain, cycling frequency, wave form of stress or strain cycle, temperature

SYNONYMS, SYMBOLS, ABBREVIATIONS, AND RELATED TERMS: Hours to Failure

Fatigue Limit

Common Abbreviation:	S_f
Preferred or SI Unit	MPa
Alternate or English Unit	ksi

DEFINITION:

The Fatigue Limit is the value of alternating stress or strain below which a material can presumably endure an infinite number of stress cycles, that is, the stress at which the S-N diagram becomes and appears to remain horizontal. For many variable-amplitude loading conditions this is true. But for conditions involving periodic overstrains, as is typical for many actual components, large changes in the long-life fatigue resistance can occur.

MATERIAL/TEST PARAMETERS: Stress ratio, surface conditions, notched or unnotched specimen

CONVERSION FACTORS:

to convert	to	multiply by	to convert	to	multiply by
MPa	ksi	0.14504	ksi	MPa	6.8948
psi	MPa	0.006895	MN/m^2	MPa	1
10^3 psi	MPa	6.895	N/mm^2	MPa	1
kgf/mm^2	MPa	9.807	tsi (short tons/in.^{2})	MPa	13.79
kgf/cm^2	MPa	980.7	tsi (long tons/in.^{2})	MPa	15.44

SYNONYMS, SYMBOLS, ABBREVIATIONS, AND RELATED TERMS: Endurance Limit, S_e

EXAMPLE VALUES:	**MPa**	**ksi**
Titanium Alloys	250-650	36-95

Fatigue Notch Factor

Acceptable Abbreviation: k_f

Preferred or SI Unit (Unitless)

DEFINITION:

The Fatigue Notch Factor (k_f) is the ratio of the Fatigue Strength of a smooth, unnotched specimen to the Fatigue Strength of a notched specimen at the same number of cycles. The Fatigue Notch Factor will vary with the life on the S-N curve and with the mean stress. At high stress levels and short cycles, the factor is usually less than at lower stress levels and longer cycles because of a reduction of the notch effect by plastic deformation.

MATERIAL/TEST PARAMETERS: Sample geometry, mean stress, stress amplitude

SYNONYMS, SYMBOLS, ABBREVIATIONS, AND RELATED TERMS: Fatigue Limit Reduction Factor, Endurance Limit Reduction Factor, Fatigue Strength Reduction Factor, K_f

Fatigue Notch Sensitivity

Common Abbreviation: q

Preferred or SI Unit (Unitless)

DEFINITION:

Fatigue Notch Sensitivity (q) is determined by comparing the Fatigue Notch Factor (k_f) and the Theoretical Stress Concentration Factor (k_t). Fatigue Notch Sensitivity is defined as $q = (k_f - 1)/(k_t - 1)$ where q may vary between 0 (indicating no notch sensitivity) and 1 (indicating full notch sensitivity). Most metals tend to become more notch sensitive at low stresses and long cycles.

MATERIAL/TEST PARAMETERS: Specimen size, stress concentrator shape and size, stress level, cycle length

Fatigue Strength

Common Abbreviation: S_n

Preferred or SI Unit: MPa
Alternate or English Unit: ksi

DEFINITION:

Fatigue Strength, which should not be confused with Fatigue Limit, is the stress to which the material can be subjected for a specified number of cycles. The term Fatigue Strength is used for materials, such as most nonferrous metals, that do not exhibit well-defined fatigue limits. Fatigue Strength is also used to describe the fatigue behavior of carbon and low-alloy steels at stresses greater than the Fatigue Limit. The Median Fatigue Strength at N cycles is an estimate of the stress levels at which 50% of the population would survive N cycles.

MATERIAL/TEST PARAMETERS: Number of Cycles

CONVERSION FACTORS:

to convert	to	multiply by	to convert	to	multiply by
MPa	ksi	0.14504	ksi	MPa	6.8948
psi	MPa	0.006895	MN/m^2	MPa	1
10^3 psi	MPa	6.895	N/mm^2	MPa	1
kgf/mm^2	MPa	9.807	tsi (short tons/in.^{2})	MPa	13.79
kgf/cm^2	MPa	980.7	tsi (long tons/in.^{2})	MPa	15.44

SYNONYMS, SYMBOLS, ABBREVIATIONS, AND RELATED TERMS: Fatigue Strength at N Cycles, Median Fatigue Strength at N Cycles

Fatigue Strength Coefficient

Common Abbreviation:	σ'_f
Preferred or SI Unit	MPa
Alternate or English Unit	ksi

DEFINITION:

The Fatigue Strength Coefficient is the true stress required to cause failure in one reversal. The coefficient is defined as the intercept at $2N_f = 1$ on the log-log plot of $2N_f$ (reversals to failure) versus $\Delta\sigma/2$ (true stress amplitude).

EXAMPLE TEST METHOD: SAE J1099

CONVERSION FACTORS:

to convert	to	multiply by	to convert	to	multiply by
MPa	ksi	0.14504	ksi	MPa	6.8948
psi	MPa	0.006895	MN/m^2	MPa	1
10^3 psi	MPa	6.895	N/mm^2	MPa	1
kgf/mm^2	MPa	9.807	tsi (short tons/in.^{2})	MPa	13.79
kgf/cm^2	MPa	980.7	tsi (long tons/in.^{2})	MPa	15.44

SYNONYMS, SYMBOLS, ABBREVIATIONS, AND RELATED TERMS: Cyclic Strength Coefficient

Fatigue Strength Exponent

Common Abbreviation: b

Preferred or SI Unit (Unitless)

DEFINITION:

The Fatigue Strength Exponent is the power to which life, in reversals to failure, must be raised to be proportional to true stress amplitude ($\Delta\sigma/2$). The exponent may be defined as the slope of the log-log plot of $2N_f$ versus $\Delta\sigma/2$.

EXAMPLE TEST METHOD: SAE J1099

SYNONYMS, SYMBOLS, ABBREVIATIONS, AND RELATED TERMS: Basquin's Exponent

Stress Concentration Factor

Common Abbreviation: k_t

Preferred or SI Unit (Unitless)

DEFINITION:

The Stress Concentration Factor is the ratio of the greatest stress in the region of a notch or other stress concentrator (as determined by the theory of elasticity or by experimental procedures) to the corresponding nominal stress.

SYNONYMS, SYMBOLS, ABBREVIATIONS, AND RELATED TERMS:
Theoretical Stress Concentration Factor, Experimental Stress Concentration Factor, K_t

Bend Radius

Common Abbreviation:	(None established)
Preferred or SI Unit	t
Alternate or English Unit	t

DEFINITION:

The Bend Radius is the smallest radius around which a specimen can be bent without cracks being observed in the outer fiber (tension) surface. Bend Radius data is almost always expressed as a multiple of the specimen thickness (t). For example, a material with a minimum bend radius of 3t can be bent without cracking through a radius equal to 3 times the specimen thickness. Bend Radius data is also commonly expressed as a ratio of the radius and the specimen thickness (r/t).

MATERIAL/TEST PARAMETERS: Specimen thickness

SYNONYMS, SYMBOLS, ABBREVIATIONS, AND RELATED TERMS: Bend Ductility Radius

Plastic Strain Ratio

Common Abbreviation: r

Preferred or SI Unit (Unitless)

DEFINITION:

The Plastic Strain Ratio (r) compares ductility in the plane and thickness directions by determining true strains (ε_w, ε_t) in the width and thickness directions, respectively, during tension tests. The ratio is expressed as $r = \varepsilon_w/\varepsilon_t$, and a higher value for r generally means a deeper draw can be achieved. Because sheet metal is anisotropic, it is necessary to use the average of the strain ratios measured parallel to, transverse to, and 45 degrees to the rolling direction of the sheet to obtain an average strain ratio, r_m, which is expressed as: $r_m = (r_L + 2r_{45} + r_T)/4$ where r_L is the strain ratio in the longitudinal direction, r_{45} is the strain ratio measured at 45 degrees to the rolling direction, and r_T is the strain ratio in the transverse direction.

EXAMPLE TEST METHODS: ASTM E 517, SAE J877

MATERIAL/TEST PARAMETERS: Percent Elongation, direction

SYNONYMS, SYMBOLS, ABBREVIATIONS, AND RELATED TERMS: Deep Drawability Index, r-value, Drawability, r_m, Anisotropy Factor

EXAMPLE VALUES:	**(Unitless)**
Carbon and Alloy Steels	0-1.6
Aluminum Alloys	0.6
Copper Alloys	0.9

Crack-Arrest Toughness

Common Abbreviation:	K_a
Preferred or SI Unit	$MPa\sqrt{m}$
Alternate or English Unit	$ksi\sqrt{in}$

DEFINITION:

Crack Arrest Toughness (K_a) is a measure of the ability of a material to stop a fast running crack. K_a is an estimate of the stress-intensity factor at a short time after crack arrest. K_{Ia} (Plane-Strain Crack-Arrest Fracture Toughness) is the value of K_a for a crack that arrests under conditions of crack-front plane-strain.

EXAMPLE TEST METHOD: ASTM E 1221

MATERIAL/TEST PARAMETERS: Test method

CONVERSION FACTORS:

to convert	to	multiply by
$MPa\sqrt{m}$	$ksi\sqrt{in}$	0.910048
$ksi\sqrt{in}$	$MPa\sqrt{m}$	1.098843
$MN\sqrt{m}$	$MPa\sqrt{m}$	1

SYNONYMS, SYMBOLS, ABBREVIATIONS, AND RELATED TERMS: K_{Ia}, Plane-Strain Crack-Arrest Fracture Toughness

Critical Value of the J-Integral

Common Abbreviation: J_{Ic}

Preferred or SI Unit: kJ/m^2
Alternate or English Unit: $in. \cdot lbf/in.^2$

DEFINITION:

The Critical Value of the J-integral (J_{Ic}) is an engineering estimate of fracture toughness near the initiation of slow stable crack growth. J_{Ic} is used for low-strength, high-toughness materials.

EXAMPLE TEST METHOD: ASTM E 813

MATERIAL/TEST PARAMETERS: Test method, loading rate, specimen geometry

CONVERSION FACTORS:

to convert	to	multiply by	to convert	to	multiply by
kJ/m^2	$in. \cdot lbf/in.^2$	5.71015	$in. \cdot lbf/in.^2$	kJ/m^2	0.175127

SYNONYMS, SYMBOLS, ABBREVIATIONS, AND RELATED TERMS: J-integral

Ductile-Brittle Transition Temperature

Common Abbreviation:	DBTT
Preferred or SI Unit	°C
Alternate or English Unit	°F

DEFINITION:

The Ductile-Brittle Transition Temperature (DBTT) is an arbitrarily defined temperature that lies within the temperature range in which metal fracture characteristics (as determined by tests of notched specimens) change rapidly. DBTT is generally used to characterize the low-temperature embrittlement of ferritic steels. The DBTT can be defined at any temperature in the region of mixed ductile and brittle fracture.

MATERIAL/TEST PARAMETERS: Criteria used (% ductile and % brittle fracture)

CONVERSION FACTORS:

to convert	to	multiply by	and then
°C	°F	1.8	add 32
°F	°C	0.55556	subtract 17.7778
K	°C	1	subtract 273.15
R (Rankine)	°C	0.55556	subtract 273.15

Dynamic Fracture Toughness

Common Abbreviation:	K_{Id}
Preferred or SI Unit	$MPa\sqrt{m}$
Alternate or English Unit	$ksi\sqrt{in}$

DEFINITION:

Dynamic Fracture Toughness (K_{Id}) is used as an approximation of Plane-Strain Fracture Toughness (K_{Ic}) for very tough materials. A quasi-static evaluation procedure can be used to determine K_{Id}, with restrictions to the input velocity depending on the toughness of the material. Alternatively, K_{Id} can be determined by performing an impact experiment and measuring the time-to-fracture. K_{Id} then equals the Dynamic Crack Tip Stress-Intensity Factor ($K_I^{dyn\,(t)}$) at the measured time-to-fracture.

MATERIAL/TEST PARAMETERS: Test method, specimen geometry, crack length

CONVERSION FACTORS:

to convert	to	multiply by
$MPa\sqrt{m}$	$ksi\sqrt{in}$	0.910048
$ksi\sqrt{in}$	$MPa\sqrt{m}$	1.098843
$MN\sqrt{m}$	$MPa\sqrt{m}$	1

SYNONYMS, SYMBOLS, ABBREVIATIONS, AND RELATED TERMS: Dynamic Critical Stress-Intensity Factor, Impact Fracture Toughness, Dynamic Impact Fracture Toughness, $K_I^{dyn\,(t)}$, Dynamic Crack Tip Stress-Intensity Factor

Dynamic Tear Energy

Common Abbreviation:	DT
Preferred or SI Unit	J
Alternate or English Unit	ft · lbf

DEFINITION:

Dynamic Tear Energy is the Impact Toughness of metals that are too tough and ductile to fracture under plane-strain conditions in sizes normally used for structures. Dynamic Tear Energy tests use larger specimens with a proportionally deeper notch than Charpy specimens and measure the total energy required to fracture dynamic tear specimens.

EXAMPLE TEST METHOD: ASTM E 604

MATERIAL/TEST PARAMETERS: Test method

CONVERSION FACTORS:

to convert	to	multiply by
J	ft · lbf	0.7375621
ft · lbf	J	1.355818
kgf m	J	9.80665
N · m	J	1

Fracture Appearance Transition Temperature

Common Abbreviation:	FATT
Preferred or SI Unit	°C
Alternate or English Unit	°F

DEFINITION:

To obtain the Fracture Appearance Transition Temperature (FATT), the percentage of shear (fibrous fracture) in a fracture surface is measured. The 50% FATT represents the temperature at 50% shear (50% fibrous).

MATERIAL/TEST PARAMETERS: Transition criteria

CONVERSION FACTORS:

to convert	to	multiply by	and then
°C	°F	1.8	add 32
°F	°C	0.55556	subtract 17.7778
K	°C	1	subtract 273.15
R (Rankine)	°C	0.55556	subtract 273.15

SYNONYMS, SYMBOLS, ABBREVIATIONS, AND RELATED TERMS: 50% FATT

Fracture Ductility

Common Abbreviation: ε_f

Preferred or SI Unit (Unitless)

DEFINITION:

Fracture Ductility is the true plastic strain at fracture, or, the ln (100/100 – %RA) where RA is the Reduction of Area. Fatigue Ductility corresponds to Fracture Ductility. Fatigue Ductility is the ability of a material to deform plastically before fracturing and is usually expressed in percent in direct analogy with elongation and reduction of area ductility measures.

EXAMPLE TEST METHOD: SAE J1099

MATERIAL/TEST PARAMETERS: Test method, gage length

SYNONYMS, SYMBOLS, ABBREVIATIONS, AND RELATED TERMS: True Fracture Ductility, Fatigue Ductility, D_f

Fracture Stress

Common Abbreviation:	S_f
Preferred or SI Unit	MPa
Alternate or English Unit	ksi

DEFINITION:

Fracture Stress can be defined as: The true normal stress on the minimum cross-sectional area at the beginning of fracture, the maximum principal true stress at fracture usually referring to unnotched tensile specimens, or the (hypothetical) true stress that will cause fracture without further deformation at any given strain. The true stress should be corrected for the triaxial state of stress existing in the tensile specimen at fracture. Because the data required for this correction frequently are not available, true fracture stress values are frequently in error.

CONVERSION FACTORS:

to convert	to	multiply by	to convert	to	multiply by
MPa	ksi	0.14504	ksi	MPa	6.8948
psi	MPa	0.006895	MN/m^2	MPa	1
10^3 psi	MPa	6.895	N/mm^2	MPa	1
kgf/mm^2	MPa	9.807	tsi (short tons/in.^{2})	MPa	13.79
kgf/cm^2	MPa	980.7	tsi (long tons/in.^{2})	MPa	15.44

SYNONYMS, SYMBOLS, ABBREVIATIONS, AND RELATED TERMS: Breaking Stress, True Fracture Stress

Fracture Toughness

Common Abbreviation:	K
Preferred or SI Unit	$MPa\sqrt{m}$
Alternate or English Unit	$ksi\sqrt{in}$

DEFINITION:

Fracture Toughness is a generic term for measures of resistance to extension of a crack. The term is sometimes restricted to results of fracture mechanics test, which are directly applicable in fracture control. However, the term commonly includes results from simple tests of notched or precracked specimens not based on fracture mechanics analysis. The Stress-Intensity Factor (K) is a scaling factor used in linear-elastic fracture mechanics to describe the intensification of applied stress at the tip of a crack of known size and shape. At the onset of rapid crack propagation in any structure containing a crack, the factor is called the Critical Stress-Intensity Factor, or the Fracture Toughness. Various subscripts are used to denote different loading conditions or Fracture Toughnesses. For example, K_0 is the value of K at the onset of rapid fracturing. K_1, K_2, K_3, K_I, K_{II}, and K_{III} are values of K at the mode denoted by the subscript. Mode 1 represents opening displacement adjacent to the crack tip, 2 represents sliding, and 3 represents tearing displacement. I and II specialize the mode to plane-strain; III to anti-plane-strain. See also Dynamic Fracture Toughness (K_{Id}), Plane-Stress Fracture Toughness (K_c), Plane-Strain Fracture Toughness (K_{Ic}), Crack Arrest Toughness (K_a), and the Critical Value of the J-Integral (J_{Ic}).

EXAMPLE TEST METHODS: ASTM D 9, ASTM E 616

MATERIAL/TEST PARAMETERS: Test method, loading conditions

CONVERSION FACTORS:

to convert	to	multiply by
$MPa\sqrt{m}$	$ksi\sqrt{in}$	0.910048
$ksi\sqrt{in}$	$MPa\sqrt{m}$	1.098843
$MN\sqrt{m}$	$MPa\sqrt{m}$	1

SYNONYMS, SYMBOLS, ABBREVIATIONS, AND RELATED TERMS: Critical Stress-Intensity Factor, Stress-Intensity Factor, k, K_0, K_1, K_2, K_3, K_I, K_{II}, K_{III}

Impact Toughness

Common Abbreviation:	(None established)
Preferred or SI Unit	J
Alternate or English Unit	ft · bf

DEFINITION:

Impact Toughness is the amount of energy required to fracture a material, usually measured as the amount of energy absorbed in breaking a Charpy or Izod test specimen under impact. Values have no intrinsic significance, but are important when empirically correlated with service experience or when used as a means of comparing materials.

EXAMPLE TEST METHOD: ASTM E 23

MATERIAL/TEST PARAMETERS: Test method (Izod or Charpy), test conditions

CONVERSION FACTORS:

to convert	to	multiply by
J	ft · lbf	0.7375621
ft · lbf	J	1.355818
kgf m	J	9.80665
N · m	J	1

SYNONYMS, SYMBOLS, ABBREVIATIONS, AND RELATED TERMS: Impact Strength, Notch Toughness, Impact Value, Impact Energy, Toughness, Impact Resistance, CVN, Charpy Impact Energy, Izod Impact Strength, Notched Bar Impact

Nil Ductility Transition Temperature

Common Abbreviation:	NDTT
Preferred or SI Unit	°C
Alternate or English Unit	°F

DEFINITION:

Nil Ductility Transition Temperature is the temperature below which steel, in the presence of a cleavage crack, will not deform plastically before fracturing, but will fracture at the moment of yielding. Just above the Nil Ductility Transition Temperature, considerable plastic deformation (bulging) precedes fracture.

EXAMPLE TEST METHOD: ASTM E 208

MATERIAL/TEST PARAMETERS: Test method

CONVERSION FACTORS:

to convert	to	multiply by	and then
°C	°F	1.8	add 32
°F	°C	0.55556	subtract 17.7778
K	°C	1	subtract 273.15
R (Rankine)	°C	0.55556	subtract 273.15

Plane-Strain Fracture Toughness

Common Abbreviation:	K_{Ic}
Preferred or SI Unit	$MPa\sqrt{m}$
Alternate or English Unit	$ksi\sqrt{in}$

DEFINITION:

Plane-Strain Fracture Toughness is the minimum value of K_c (the plane-stress Fracture Toughness) for any given condition that is attained when rapid crack propagation in the opening mode is governed by plane-strain conditions.

EXAMPLE TEST METHOD: ASTM E 399

MATERIAL/TEST PARAMETERS: Orientation, temperature, test method, test conditions

CONVERSION FACTORS:

to convert	to	multiply by
$MPa\sqrt{m}$	$ksi\sqrt{in}$	0.910048
$ksi\sqrt{in}$	$MPa\sqrt{m}$	1.098843
$MN\sqrt{m}$	$MPa\sqrt{m}$	1

Plane-Stress Fracture Toughness

Common Abbreviation:	K_c
Preferred or SI Unit	$MPa\sqrt{m}$
Alternate or English Unit	$ksi\sqrt{in}$

DEFINITION:

Plane-Stress Fracture Toughness is the value of stress intensity at which crack propagation becomes rapid in sections thinner than those in which plane-strain conditions prevail.

CONVERSION FACTORS:

to convert	to	multiply by
$MPa\sqrt{m}$	$ksi\sqrt{in}$	0.910048
$ksi\sqrt{in}$	$MPa\sqrt{m}$	1.098843
$MN\sqrt{m}$	$MPa\sqrt{m}$	1

Table 1. Hardness Conversions and Typical Ranges

Note: The physical properties of the metal affect conversion to some extent. (Numbers in parentheses are outside the range of effectiveness of the test)

Vickers Hardness No. (VHN)	50	100	150	200	250	300	350	400	450	500	550	600	650	700	750	800	850	900
Knoop Hardness No.		112	164	216	262	309	356	412	471	528	583	636	687	735	780	822	858	895
Brinell, 500 kg (HB)	51	90	125	160	197													
Brinell, 3000 kg std. ball (HB)	47	101	143	190	239	284	331	380	425	(465)	(505)							
Brinell, 3000 kg, WC ball (HB)	47	101	143	190	239	284	322	380	425	471	517	564	611	(656)	(691)	(722)	(751)	(777)
Rockwell A scale (HRA)		38	44	56	62	65	68	71	73	75	77	79	80	81	82	83	84	85
Rockwell B scale (HRB)		56	77	92	100	(106)	(109)											
Rockwell C scale (HRC)				(11)	32	30	36	41	45	49	52	55	58	60	62	64	66	67
Scleroscope Hardness			22	29	37	45	51	55	59	66	70	74	77	81	86	88	91	
Steel alloys, Annealed		■	■	■	■	■												
Steel alloys, Hardened						■	■	■	■	■	■	■	■	■	■			
Aluminum alloys, Non-heat treatable	■	■																
Aluminum Alloys, Heat treatable		■	■															
Copper alloys	■	■	■	■	■		Beryllium	Coppers										
Stainless steel, Austenitic, Annealed		■																
Stainless steel, Martensitic						■	■	■	■	■	■	■	■	■	■			
Magnesium alloys	■	■																
Titanium alloys			■	■	■	■	■	■										

Table 2. Rockwell Hardness Scales, Load and Indentors

Scale designation	Indentor Type	Indentor Diam, in.	Major load, kg	Scale designation	Indentor Type	Indentor Diam, in.	Major load, kg
Regular Rockwell tester				**Superficial Rockwell tester**			
A	Brale	. . .	60	15N	N Brale	. . .	15
B	Ball	1/16	100	45N	N Brale	. . .	45
C	Brale	. . .	150	30N	N Brale	. . .	30
D	Brale	. . .	100	15T	Ball	1/16	15
E	Ball	1/8	100	30T	Ball	1/16	30
F	Ball	1/16	60	45T	Ball	1/16	45
G	Ball	1/16	150	15W	Ball	1/8	15
H	Ball	1/8	60	30W	Ball	1/8	30
K	Ball	1/8	150	45W	Ball	1/8	45
L	Ball	¼	60	15X	Ball	1/4	15
M	Ball	¼	100	30X	Ball	1/4	30
P	Ball	¼	150	45X	Ball	1/4	45
R	Ball	½	60	15Y	Ball	1/2	15
S	Ball	½	100	30Y	Ball	1/2	30
V	Ball	½	150	45Y	Ball	1/2	45

Brinell Hardness

Common Abbreviation: HB

Preferred or SI Unit (Hardness number)

DEFINITION:

Hardness is a measure of the resistance of a material to surface indentation or abrasion, one or a measure of a material's elastic properties. Hardness may be thought of as a function of the stress required to produce some specified type of surface deformation. A Brinell Hardness number (HB) is the result of a static indentation test in which a hardened steel or tungsten carbide ball is pressed into the sample surface and the size of the indentation is measured optically. HB is the ratio of the applied force to the surface area of the impression in units of kgf/mm^2 as follows:

$$HB = \frac{2P}{\pi D(D - \sqrt{D^2 - d^2})}$$

where P is load, kg; D is ball diameter, mm; and d is diameter of the indentation, mm. Note that a multiplier of 0.102 needs to be applied if the test force is given in newtons. Note also that a multiplier of 9.81 needs to be used to convert from the kgf/mm^2 base set to MPa. The standard ball is 10 mm in diameter, and the applied force varies from 500 to 3000 kg in 500 kg increments. A 500 kg force is generally used for most nonferrous metals. A 3000 kg force is used for ferrous metals and some higher-strength nonferrous metals such as cobalt, nickel, and high-strength titanium alloys. Carbide indentors are used for materials harder than HB 444 due to flattening of the steel balls. The large surface area makes Brinell hardness suitable for materials with coarse microstructures such as cast iron. Brinell is not suitable for case-hardened steels or thin surface coatings that will cave in under the applied load.

EXAMPLE TEST METHOD: ASTM E 10

MATERIAL/TEST PARAMETERS: Test equipment: ball size, ball type, loading force, spacing between repeat impressions. Specimen: surface finish, surface flatness, thickness, parallelism of test and supported surfaces.

CONVERSION FACTORS:

See Hardness Table 1, Page 65

SYNONYMS, SYMBOLS, ABBREVIATIONS, AND RELATED TERMS: HBN, BHN, Bhn, Brinell Hardness Number, HBS (steel ball), HBW (tungsten carbide ball)

EXAMPLE VALUES:

See Hardness Table 1, Page 65

Knoop Hardness

Common Abbreviation: HK

Preferred or SI Unit (Hardness number)

DEFINITION:

Hardness is a measure of the resistance of a material to surface indentation or abrasion, or a measure of a material's elastic properties. Hardness may be thought of as a function of the stress required to produce some specified type of surface deformation. A Knoop hardness number (HK) is the result of a static indentation test in which a diamond indentor is pressed into a sample surface, followed by a measurement of the length of the indentation. HK is a number related to the applied load (in kgf) and to the projected area (in mm^2) of the permanent impression made by a rhombic-based pyramidal diamond indentor having specified angle edges. Test loads are between 1 and 1000 gm.

EXAMPLE TEST METHOD: ASTM E 384

MATERIAL/TEST PARAMETERS: Test equipment: loading force, indentor edge angle

CONVERSION FACTORS:

See Hardness Table 1, Page 65

SYNONYMS, SYMBOLS, ABBREVIATIONS, AND RELATED TERMS:
Microhardness

EXAMPLE VALUES:

See Hardness Table 1, Page 65

Rockwell Hardness

Common Abbreviation: HRx (x = one of 30 possible scales; see Table 2, Page 65)

Preferred or SI Unit (Hardness number)

DEFINITION:

Hardness is a measure of the resistance of a material to surface indentation or abrasion, or a measure of a material's elastic properties. Hardness may be thought of as a function of the stress required to produce some specified type of surface deformation. A Rockwell hardness number (HRX) is the result of a static indentation test in which a hardened steel ball or a 120-degree conical diamond indentor is pressed into a sample surface, followed the measurement of the depth of the indentation. HRX is determined directly by the test equipment as a function of the depth of the impression, applied load, and type of indentor. Diamond indentors are used for ferrous metals and some higher-strength nonferrous metals such as cobalt, nickel, and high-strength titanium alloys. Various sizes of steel balls are used with different loads for softer materials. Rockwell hardness testing is the most widely used method for determining hardness because it is simple to perform and does not require highly skilled operators. The primary scales used are HRB for nonferrous alloys and annealed steels, and HRC for the higher-strength alloys or quenched-and-tempered steels. The other scales cover the full range of metals from soft solders to case-hardened steels. See Hardness Table 2, Page 65, for a complete list of Rockwell scale designations, loads, and indentors.

EXAMPLE TEST METHOD: ASTM E 18

MATERIAL/TEST PARAMETERS: Test equipment: scale, indentor type, loading force, spacing between repeat impressions. Specimen: material type, surface finish, surface flatness, thickness, parallelism of test and supported surfaces

CONVERSION FACTORS:

See Hardness Table 1, Page 65

SYNONYMS, SYMBOLS, ABBREVIATIONS, AND RELATED TERMS: HRA, HRB, HRC, HRD, HRE, HRF, HRG, HRH, HRK, HRL, HRM, HRP, HRR, HRS, HRV, Rock A, Rock B, Rock C, HR15N, HR30N, HR45N, HR15T, HR30T, HR45T, HR15W, HR30W, HR45W, HR15X, HR30S, HR45X, HR15Y, HR30Y, HR45Y, Rockwell Superficial Hardness

EXAMPLE VALUES:

See Hardness Table 1, Page 65

Scleroscope Hardness

Common Abbreviation: HSc

Preferred or SI Unit (Hardness number)

DEFINITION:

Hardness is a measure of the resistance of a material to surface indentation or abrasion, or a measure of material's elastic properties. Hardness may be thought of as a function of the stress required to produce some specified type of surface deformation. Scleroscope hardness is a number related to the height of rebound of a diamond-tipped hammer, called a tup, that is dropped on the material being tested, with the rebound height indicative of the elasticity. The scale is graduated in "Shore" units.

EXAMPLE TEST METHOD: ASTM E 448

CONVERSION FACTORS:

See Hardness Table 1, Page 65

SYNONYMS, SYMBOLS, ABBREVIATIONS, AND RELATED TERMS: HSd, Shore Hardness

EXAMPLE VALUES:

See Hardness Table 1, Page 65

Vickers Hardness

Common Abbreviation: HV

Preferred or SI Unit (Hardness number)

DEFINITION:

Hardness is a measure of the resistance of a material to surface indentation or abrasion, or a measure of a material's elastic properties. Hardness may be thought of as a function of the stress required to produce some specified type of surface deformation. A Vickers hardness number (HV) is the result of a Vickers hardness test and is related to the applied load (in kgf) and to the surface area (in mm^2) of the permanent impression made by a square-based pyramidal diamond indentor having included face angles of 136 degrees. The test is useful for a wide range of hardness from very soft to very hard.

EXAMPLE TEST METHOD: ASTM E 92

CONVERSION FACTORS:

See Hardness Table 1, Page 65

SYNONYMS, SYMBOLS, ABBREVIATIONS, AND RELATED TERMS: Diamond Pyramid Hardness, Diamond Pyramid Number, HY, DPN, VPN

EXAMPLE VALUES:

See Hardness Table 1, Page 65

Shear Strength

Common Abbreviation:	USS
Preferred or SI Unit	MPa
Alternate or English Unit	ksi

DEFINITION:

Shear Strength (USS) is the maximum shear stress that a material is capable of sustaining. USS is calculated from the maximum load during a shear or torsion test and is based on the original dimensions of the cross section of the specimen. F_{su} is an abbreviation used for allowable ultimate stress in pure shear, representing the average shearing stress over the cross section.

MATERIAL/TEST PARAMETERS: Form, temper, sample size, temperature, test method (torsion, blanking or double (rivet) shear)

CONVERSION FACTORS:

to convert	to	multiply by	to convert	to	multiply by
MPa	ksi	0.14504	ksi	MPa	6.8948
psi	MPa	0.006895	MN/m^2	MPa	1
10^3 psi	MPa	6.895	N/mm^2	MPa	1
kgf/mm^2	MPa	9.807	tsi (short tons/in.^{2})	MPa	13.79
kgf/cm^2	MPa	980.7	tsi (long tons/in.^{2})	MPa	15.44

SYNONYMS, SYMBOLS, ABBREVIATIONS, AND RELATED TERMS: S_{us}, Ultimate Shear Stress, USS, Apparent Ultimate Shear Strength, Modulus of Rupture, Modulus of Rupture in Torsion, F_{su}

EXAMPLE VALUES:	**MPa**	**ksi**
Aluminum Alloys	70-280	10-40
Titanium Alloys	400-680	60-99

Shear Yield Strength

Common Abbreviation:	SYS
Preferred or SI Unit	MPa
Alternate or English Unit	ksi

DEFINITION:

Shear Yield Strength (SYS) is the maximum shear stress that can be applied without causing permanent deformation or exceeding a specified deviation from stress-strain proportionality. It is typically calculated from the torque at yielding value in a torsion test and the dimensions of the torsion test specimen. Shear yield strength is generally defined as the maximum stress developed by a torque producing an offset of 0.2% from the original modulus line (a method analogous to that used for determining tensile yield strength).

MATERIAL/TEST PARAMETERS: Offset, form, temper, sample size, temperature, test method (torsion, blanking, or double (rivet) shear)

CONVERSION FACTORS:

to convert	to	multiply by	to convert	to	multiply by
MPa	ksi	0.14504	ksi	MPa	6.8948
psi	MPa	0.006895	MN/m^2	MPa	1
10^3 psi	MPa	6.895	N/mm^2	MPa	1
kgf/mm^2	MPa	9.807	tsi (short tons/in.^{2})	MPa	13.79
kgf/cm^2	MPa	980.7	tsi (long tons/in.^{2})	MPa	15.44

SYNONYMS, SYMBOLS, ABBREVIATIONS, AND RELATED TERMS: S_{sy}, Shearing Yield Strength, Shear Yield Stress, F_{sy}, Shear Proof Strength

Elongation

Common Abbreviation:	El
Preferred or SI Unit	% (of gage length)
Alternate or English Unit	% (of gage length)

DEFINITION:

Tensile Elongation is a measure of ductility, or the ability of a material to deform plastically before fracturing. Elongation (El) is the increase in a predefined standard gage length of a tension test specimen and is usually expressed as a percentage of the original gage length. The increase in gage length may be determined either at fracture (Maximum Elongation, Ultimate Elongation, or Break Elongation), or after fracture (Total Elongation or El_t), as specified for the material under test. The term elongation, when applied to metals, generally means measurement after fracture. Uniform Elongation (El_u or e_u) is the elongation at maximum load and immediately preceding the onset of necking in a tensile test. Yield Point Elongation (YPE) is the difference between the elongation at the completion and at the start of discontinuous yielding and is applicable to materials that exhibit a Yield Point.

EXAMPLE TEST METHOD: ASTM E 8

MATERIAL/TEST PARAMETERS: Gage length, sample size, test procedure

SYNONYMS, SYMBOLS, ABBREVIATIONS, AND RELATED TERMS:
Elongation at Break, Elongation at Fracture, Ductility, A, Allongement, Dehnung, Break Elongation, El_t, Total Elongation, Maximum Elongation, Uniform Elongation, e, e_f, e_u, El_u, Ultimate Elongation, YPE, Yield Point Elongation

EXAMPLE VALUES:	**% (of gage length)**
Carbon and Alloy Steels	5-40
Aluminum Alloys	1-40
Copper Alloys	1-70
Magnesium Alloys	4-16
Stainless Steels	15-60
Titanium Alloys	4-24

Notch Tensile Strength

Common Abbreviation:	NTS
Preferred or SI Unit	MPa
Alternate or English Unit	ksi

DEFINITION:

Notch Tensile Strength (NTS) is defined as the maximum nominal (net-section) stress that a notched tensile specimen is capable of sustaining. NTS is the maximum load on a notched tensile test specimen, divided by the minimum cross-sectional area (the area at the root of the root of the notch). Because of the plastic constraint at the notch, this value will be higher than the Tensile Strength of an unnotched specimen if the material possesses some ductility. A common way of detecting notch brittleness (or high notch sensitivity) is by determining the Notch Strength Ratio (NSR). The NSR is the NTS divided by the Tensile Strength for an unnotched specimen. If the NSR is less than unity, the material is said to be notch brittle.

MATERIAL/TEST PARAMETERS: Form, temper, sample size, temperature, notch size and type, theoretical stress concentration factor

CONVERSION FACTORS:

to convert	to	multiply by	to convert	to	multiply by
MPa	ksi	0.14504	ksi	MPa	6.8948
psi	MPa	0.006895	MN/m^2	MPa	1
10^3 psi	MPa	6.895	N/mm^2	MPa	1
kgf/mm^2	MPa	9.807	tsi (short tons/in.^{2})	MPa	13.79
kgf/cm^2	MPa	980.7	tsi (long tons/in.^{2})	MPa	15.44

SYNONYMS, SYMBOLS, ABBREVIATIONS, AND RELATED TERMS: Notch Strength, NSR, Notch Strength Ratio, Notch Brittleness

Reduction of Area

Common Abbreviation:	RA
Preferred or SI Unit	%
Alternate or English Unit	%

DEFINITION:

Reduction of Area (RA) is a measure of ductility, or the ability of a material to deform plastically before fracturing. To determine RA the difference between the original cross-sectional area of a tension test specimen and the area of its smallest cross section is determined. RA is that difference expressed as a percentage of the original cross-sectional area of the specimen. The smallest cross section may be measured at or after fracture as specified for the materials under test. When applied to metals, RA usually means measurement after fracture.

EXAMPLE TEST METHOD: ASTM E 8

MATERIAL/TEST PARAMETERS: Specimen dimensions, temperature, test method

SYNONYMS, SYMBOLS, ABBREVIATIONS, AND RELATED TERMS: Area Reduction, Tensile Reduction in Area, Reduction in Area, Ductility

EXAMPLE VALUES:	**%**
Carbon and Alloy Steels	27-67
Titanium Alloys	12-58

Tensile Strength

Common Abbreviation:	UTS
Preferred or SI Unit	MPa
Alternate or English Unit	ksi

DEFINITION:

Tensile strength (UTS) is the maximum tensile stress that a material is capable of sustaining. UTS is calculated from the maximum load during a tension test carried to rupture and the original cross-sectional area of the specimen. F_{tu} is an abbreviation for allowable Tensile Stress.

EXAMPLE TEST METHODS: ASTM E 8, ASTM E 345, ASTM E 1450

MATERIAL/TEST PARAMETERS: Form, temper, sample size, temperature, test method

CONVERSION FACTORS:

to convert	to	multiply by	to convert	to	multiply by
MPa	ksi	0.14504	ksi	MPa	6.8948
psi	MPa	0.006895	MN/m^2	MPa	1
10^3 psi	MPa	6.895	N/mm^2	MPa	1
kgf/mm^2	MPa	9.807	tsi (short tons/in.^{2})	MPa	13.79
kgf/cm^2	MPa	980.7	tsi (long tons/in.^{2})	MPa	15.44

SYNONYMS, SYMBOLS, ABBREVIATIONS, AND RELATED TERMS: Ultimate Tensile Stress, Ultimate Strength, Maximum Tensile Strength, Tenacity, US, S_{ut}, F_{tu}, TS, S_{tu}, R_m, Resistance a la Traction, Zugfestigkeit, Engineering Tensile Strength, Tensile Stress, S_u

EXAMPLE VALUES:	**MPa**	**ksi**
Carbon and Alloy Steels	100-2300	15-335
Aluminum Alloys	70-700	10-100
Copper Alloys	170-1500	25-220
Stainless Steels	400-1500	60-220
Titanium Alloys	240-1500	35-215

Tensile Yield Strength

Common Abbreviation:	YS
Preferred or SI Unit	MPa
Alternate or English Unit	ksi

DEFINITION:

Tensile Yield Strength (YS) is the stress at which a material exhibits what is essentially the beginning point of design failure, that is, a specified deviation from proportionality of stress and strain, or a deviation from Hooke's law. The deviation criterion is typically a 0.2% offset when a material has no definite Yield Point. Tensile Yield Strengths are affected by previous compressive strain. F_{ty} is an abbreviation used for allowable Tensile Yield Stress at which permanent strain equals 0.002.

EXAMPLE TEST METHODS: ASTM E 8, ASTM E 345, ASTM E 1450

MATERIAL/TEST PARAMETERS: Test method, loading rate, offset, sample size, temperature, form, temper

CONVERSION FACTORS:

to convert	to	multiply by	to convert	to	multiply by
MPa	ksi	0.14504	ksi	MPa	6.8948
psi	MPa	0.006895	MN/m^2	MPa	1
10^3 psi	MPa	6.895	N/mm^2	MPa	1
kgf/mm^2	MPa	9.807	tsi (short tons/in.^{2})	MPa	13.79
kgf/cm^2	MPa	980.7	tsi (long tons/in.^{2})	MPa	15.44

SYNONYMS, SYMBOLS, ABBREVIATIONS, AND RELATED TERMS: Yield Stress, YS, Tensile Yield Stress, Tensile Strength at Yield, Proof Stress, Proof Strength. S_{ty}, R_e, Limite d'Elasticite, Streckgrenze, F_{ty}

EXAMPLE VALUES:	**MPa**	**ksi**
Stainless Steels	70-1050	10-150

Yield Point

Common Abbreviation:	YP
Preferred or SI Unit	MPa
Alternate or English Unit	ksi

DEFINITION:

In a test in which stresses and strains are determined for a material that exhibits the phenomenon of discontinuous yielding, the Yield Point is the first engineering stress at which an increase in strain occurs without an increase in stress, that is, essentially the beginning point of design failure. For materials that do not exhibit a Yield Point, Yield Strength serves the same purpose as Yield Point. The Elastic Limit is the maximum stress that a material is capable of sustaining without any permanent strain (deformation) remaining upon complete release of the stress. The Proportional Limit is the greatest stress a material is capable of developing without a deviation from straight-line proportionality between stress and strain.

MATERIAL/TEST PARAMETERS: Test method

CONVERSION FACTORS:

to convert	to	multiply by	to convert	to	multiply by
MPa	ksi	0.14504	ksi	MPa	6.8948
psi	MPa	0.006895	MN/m^2	MPa	1
10^3 psi	MPa	6.895	N/mm^2	MPa	1
kgf/mm^2	MPa	9.807	tsi (short tons/in.^{2})	MPa	13.79
kgf/cm^2	MPa	980.7	tsi (long tons/in.^{2})	MPa	15.44

SYNONYMS, SYMBOLS, ABBREVIATIONS, AND RELATED TERMS: Upper Yield Point, Lower Yield Point, Proportional Limit, Elastic Limit

Atomic Number

Common Abbreviation:	Z
Preferred or SI Unit	(Unitless)

DEFINITION:

The Atomic Number (Z) is the number of elementary positive charges (protons) contained within the nucleus of an atom. For an electrically neutral atom, the number of planetary electrons is also given by the Atomic Number.

SYNONYMS, SYMBOLS, ABBREVIATIONS, AND RELATED TERMS: Nuclear Charge

EXAMPLE VALUES:	**(Unitless)**
Iron	26
Aluminum	13
Copper	29
Magnesium	12
Titanium	22

Atomic Weight

Common Abbreviation:	A_r
Preferred or SI Unit	Atomic Mass Unit

DEFINITION:

Atomic Weight is the average mass of a single atom of an element, based on a relative scale in which Carbon 12 has an assigned value of 12 Atomic Mass Units. For example, 12 is the Atomic Weight of Carbon 12.

SYNONYMS, SYMBOLS, ABBREVIATIONS, AND RELATED TERMS: Atomic Mass, Relative Atomic Mass

EXAMPLE VALUES:	**Atomic Mass Unit**
Iron	55.847
Aluminum	26.981539
Copper	63.546
Magnesium	24.305
Titanium	47.88

Ionization Potential

Common Abbreviation:	(None established)
Preferred or SI Unit	V

DEFINITION:

Ionization Potential is the minimum energy per unit charge needed to remove an electron from its orbit. The potential required to remove the first electron from a neutral atom is called the First Ionization Potential.

SYNONYMS, SYMBOLS, ABBREVIATIONS, AND RELATED TERMS: Ion Potential, First Ionization Potential

EXAMPLE VALUES:	**V**
Iron	7.9024
Aluminum	5.98577
Copper	7.72638
Magnesium	7.64624
Titanium	6.8282

Neutron Cross Section

Common Abbreviation:	(None established)
Preferred or SI Unit	barn
Alternate or English Unit	cm^2

DEFINITION:

The Neutron Cross Section is a measure of the probability that an interaction of a given kind will take place between a nucleus and an incident neutron; it is an area such that the number of interactions that occur in a sample exposed to a beam of neutrons is equal to the product of the cross section, the number of nuclei per unit volume in the sample, the thickness of the sample, and the number of neutrons in the beam that would enter the sample if their velocities were perpendicular to it.

CONVERSION FACTORS:

to convert	to	multiply by
barn	cm^2	10^{-24}
cm^2	barn	10^{24}
mbarn	barn	0.001

SYNONYMS, SYMBOLS, ABBREVIATIONS, AND RELATED TERMS: Neutron Capture Cross Section, Thermal Neutron Cross Section

EXAMPLE VALUES:	**barn**	**cm^2**
Iron	2.56	2.56×10^{-24}
Aluminum	0.232	2.32×10^{-25}
Copper	3.8	3.8×10^{-24}
Magnesium	0.064	6.4×10^{-26}
Titanium	6.1	6.1×10^{-24}

Corrosion Potential

Common Abbreviation:	E_{corr}
Preferred or SI Unit	V
Alternate or English Unit	mV

DEFINITION:
Corrosion Potential refers to the potential of a corroding surface in an electrolyte, relative to a reference electrode under open-circuit conditions. SCE potential is measured against a saturated calomel electrode (SCE) composed of mercury, mercurous chloride (calomel), and a saturated chloride solution. SHE potential is measured against a standard hydrogen electrode (SHE) constructed from an inert metal electrode, usually platinum, and a solution with a unit activity of hydrogen ions (pH = 1), and saturated with hydrogen gas at 1 atm pressure. The copper sulfate/copper ($CuSO_4$/Cu) electrode is a commonly used reference electrode for field work. It contains a copper electrode in a saturated copper sulfate solution.

EXAMPLE TEST METHODS: ASTM G 3, ASTM G 5, ASTM G 59, ASTM G 61

MATERIAL/TEST PARAMETERS: Reference electrode type, electrolyte composition and concentration, partial pressure, temperature, test environment

CONVERSION FACTORS:

to convert	to	multiply by
V	mV	1,000
mV	V	0.001

to convert	to	multiply by
V vs. SHE	V vs. SCE	subtract 0.24 V
V vs. SHE	V vs. $CuSO_4$/Cu	subtract 0.31 V
V vs. SCE	V vs. $CuSO_4$/Cu	subtract 0.07 V

SYNONYMS, SYMBOLS, ABBREVIATIONS, AND RELATED TERMS: Rest Potential, Open-Circuit Potential, Freely Corroding Potential, OCP, Free Corrosion Potential, SCE Potential, SHE Potential

EXAMPLE VALUES:	**V vs. SCE**	**Environment**
Carbon and Alloy Steels	–0.7 to –0.6	Seawater
Aluminum Alloys	–0.8 to 1.0; –0.6	Seawater; Horse serum
Copper Alloys	–0.4 to –0.2; –0.1	Seawater; Horse serum
Magnesium Alloys	–1.6; –1.6	Seawater; Horse serum
Stainless Steels (Austenitic)	–0.5 to 0; 0.5	Seawater; Horse serum
Titanium Alloys	0 to 0.5; 3.5	Seawater; Horse serum

Corrosion Rate

Common Abbreviation:	(None established)
Preferred or SI Unit	mm/yr
Alternate or English Unit	mils/yr

DEFINITION:

Corrosion Rate is a measurement of the corrosion effect on a metal per unit of time. The type of corrosion rate used depends on the technical system and on the type of corrosion effect. Thus, Corrosion Rate may be expressed as an increase in corrosion depth per unit of time (Penetration Rate) or as the mass of metal turned into corrosion products per unit area of surface per unit of time (see Weight Loss and Weight Gain). The corrosion effect may vary with time and may not be the same at all points of the corroding surface.

EXAMPLE TEST METHODS: ASTM G 50, ASTM G 52

MATERIAL/TEST PARAMETERS: Test method, environment, concentration, duration of test

CONVERSION FACTORS:

to convert	to	multiply by
mm/yr	mils/yr	39.37
mils/yr	mm/yr	0.0254
in./yr	mm/yr	2.54
nm/yr	mm/yr	1,000,000
µm/yr	mm/yr	1,000

SYNONYMS, SYMBOLS, ABBREVIATIONS, AND RELATED TERMS:
Penetration Rate

EXAMPLE VALUES:	**mm/yr**	**mils/yr**
Carbon and Alloy Steels	Seawater 0.03-2.6	Seawater 1.3-101
Aluminum Alloys	Seawater 0.0009-0.28	Seawater 0.036-11
Copper Alloys	Seawater 0.002-1.0	Seawater 0.08-40
Stainless Steels	Seawater 0.003-1.9	Seawater 0.1-75
Titanium Alloys	Seawater 0.001-0.003	Seawater 0.01-0.1

Critical Pitting Potential

Common Abbreviation: CPT

Preferred or SI Unit V

DEFINITION:

The Critical Pitting Potential refers to the lowest oxidizing potential at which pits nucleate and grow.

MATERIAL/TEST PARAMETERS: Test method, electrode type

SYNONYMS, SYMBOLS, ABBREVIATIONS, AND RELATED TERMS: E_{cp}, E_p, E_{pp}

Decomposition Potential

Common Abbreviation: (None established)

Preferred or SI Unit V

DEFINITION:

Decomposition Potential refers to the Surface Potential necessary to decompose the electrolyte of a cell.

MATERIAL/TEST PARAMETERS: Test method, electrode type

CONVERSION FACTORS:

to convert	to	multiply by
V	mV	1000
mV	V	0.001
V vs. SHE	V vs. SCE	subtract 0.24 V
V vs. SHE	V vs. $CuSO_4$/Cu	subtract 0.31 V
V vs. SCE	V vs. $CuSO_4$/Cu	subtract 0.07 V

SYNONYMS, SYMBOLS, ABBREVIATIONS, AND RELATED TERMS: Surface Potential

Electrode Potential

Common Abbreviation:	EP
Preferred or SI Unit	V

DEFINITION:

Electrode Potential refers to the voltage of an electrode in an electrolyte as measured against a reference electrode under open-circuit (no current) conditions. It represents the reversible work to move a unit charge from the electrode surface through the solution to the reference electrode. The Electrode Potential does not include any resistance losses in potential in either the solution or external circuit.

MATERIAL/TEST PARAMETERS: Electrolyte composition and concentration, partial pressure and temperature, reference electrode type

CONVERSION FACTORS:

to convert	to	multiply by
V	mV	1,000
mV	V	0.001
V vs. SHE	V vs. SCE	subtract 0.24 V
V vs. SHE	V vs. $CuSO_4$/Cu	subtract 0.31 V
V vs. SCE	V vs. $CuSO_4$/Cu	subtract 0.07 V

SYNONYMS, SYMBOLS, ABBREVIATIONS, AND RELATED TERMS: Steady-State Electrode Potential

EXAMPLE VALUES:	**V vs. SCE**
Iron	–0.68
Aluminum	–1.9
Copper	+0.28
Magnesium	–2.61
Titanium	–1.91

Electrokinetic Potential

Common Abbreviation: (None established)

Preferred or SI Unit V

DEFINITION:

The Electrokinetic Potential is a potential difference in a solution caused by residual, unbalanced charge distribution in an adjoining solution, producing a double layer. The Electrokinetic Potential is different from the Electrode Potential in that it occurs exclusively in the solution phase; that is, it represents the reversible work necessary to bring a unit charge from infinity in the solution up to the interface in question, but not through the interface.

SYNONYMS, SYMBOLS, ABBREVIATIONS, AND RELATED TERMS: Zeta Potential

Equilibrium Potential

Common Abbreviation: (None established)

Preferred or SI Unit V

DEFINITION:

The Equilibrium (reversible) Potential is the voltage of an electrode in an electrolytic solution when the forward rate of a given reaction is exactly equal to the reverse rate. The Equilibrium Potential can only be defined with respect to a specific electrochemical reaction.

MATERIAL/TEST PARAMETERS: Specific reaction

SYNONYMS, SYMBOLS, ABBREVIATIONS, AND RELATED TERMS: Reversible Potential

Redox Potential

Common Abbreviation:	(None established)
Preferred or SI Unit	V

DEFINITION:

Redox Potential is the voltage of a reversible oxidation-reduction electrode measured with respect to a reference electrode (corrected to the hydrogen electrode) in an electrolyte.

EXAMPLE VALUES:	**V vs. SHE**
$Fe \leftrightarrow Fe^{2+}$	–0.440
$Al \leftrightarrow Al^{3+}$	–1.66
$Cu \leftrightarrow Cu^{2+}$	0.377
$Mg \leftrightarrow Mg^{2+}$	–2.36
$Ti \leftrightarrow Ti^{2+}$	–1.63

Repassivation Potential

Common Abbreviation: E_p

Preferred or SI Unit V

DEFINITION:

A Repassivation Potential is the minimum voltage below which pitting cannot be sustained.

SCC Threshold

Common Abbreviation:	K_{th}
Preferred or SI Unit	MPa $\sqrt{m}$
Alternate or English Unit	ksi $\sqrt{ksi}$

DEFINITION:

The SCC (Stress Corrosion Cracking) Threshold is a value of stress intensity that is characteristic of a specific combination of material, material condition, and corrosive environment above which stress corrosion crack propagation occurs, and below which the material is immune from stress corrosion cracking. The threshold stress intensity for stress corrosion cracking is denoted K_{ISCC} when loading conditions meet plane-strain requirements.

MATERIAL/TEST PARAMETERS: Loading conditions

CONVERSION FACTORS:

to convert	to	multiply by
MPa $\sqrt{m}$	ksi $\sqrt{ksi}$	0.910048
ksi $\sqrt{ksi}$	MPa $\sqrt{m}$	1.098843

SYNONYMS, SYMBOLS, ABBREVIATIONS, AND RELATED TERMS: K_{ISCC}, Threshold Stress Intensity, Stress Corrosion Cracking Threshold

Standard Electrode Potential

Common Abbreviation: (None established)

Preferred or SI Unit V

DEFINITION:

The Standard Electrode Potential is the reversible potential for an electrode process when all products and reactions are at unit activity on a scale in which the potential for the standard hydrogen half-cell is zero. Standard Electrode Potential is applicable only to reference electrodes or to laboratory conditions with pure metals at constant temperature, concentration, and pressure. Under field conditions, only the reference electrode will be at or near the standard electrode potential, while the other metal will be in a natural or test solution in which its electrode potential may be significantly different from its standard electrode potential. The usual standard electrodes are: standard hydrogen electrode (SHE), standard calomel electrode (SCE), and copper sulfate/copper electrode ($CuSO_4$/Cu). Standard conditions in both half-cells include one molar concentrations of the respective metal ions, standard temperature, standard pressure, and equilibrium conditions.

MATERIAL/TEST PARAMETERS: Electrolyte concentration, temperature, pressure

CONVERSION FACTORS:

to convert	to	multiply by
V	mV	1,000
mV	V	0.001
V vs. SHE	V vs. SCE	subtract 0.24 V
V vs. SHE	V vs. $CuSO_4$/Cu	subtract 0.31 V
V vs. SCE	V vs. $CuSO_4$/Cu	subtract 0.07 V

EXAMPLE VALUES:	**V vs. SHE**
Iron	–0.44
Aluminum	–1.66
Copper	+0.52
Magnesium	–2.37
Titanium	–1.63

Weight Gain

Common Abbreviation: (None established)

Preferred or SI Unit mg/cm^2

DEFINITION:

Weight Gain is an increase in a specimen's weight after exposure to a corrosive environment, often due to a buildup of corrosion products.

SYNONYMS, SYMBOLS, ABBREVIATIONS, AND RELATED TERMS: Corrosion Rate (Weight Gain)

Weight Loss

Common Abbreviation: (None established)

Preferred or SI Unit mg/cm^2

DEFINITION:

Weight Loss is a decrease in specimen weight after exposure to a corrosive environment.

SYNONYMS, SYMBOLS, ABBREVIATIONS, AND RELATED TERMS: Corrosion Rate (Weight Loss)

Critical Current Density of Superconductivity

Common Abbreviation:	J_c
Preferred or SI Unit	A/mm^2
Alternate or English Unit	$A/in.^2$

DEFINITION:

The Critical Current Density of Superconductivity (J_c) is the critical current (I_c, or the maximum electrical current below which a superconductor exhibits superconductivity at some given temperature and magnetic field) divided by the total cross-sectional area of the conductor. J_c is one of the three upper thresholds that define the operating region for a stable state of superconductivity. The other thresholds are the Critical Field Strength of Superconductivity and the Critical Temperature of Superconductivity. The three critical parameters of superconductivity are closely interdependent. For example, J_c decreases with increases in field strength or temperature.

EXAMPLE TEST METHOD: ASTM B 713

MATERIAL/TEST PARAMETERS: Magnetic field strength, temperature

CONVERSION FACTORS:

to convert	to	multiply by
A/mm^2	$A/in.^2$	645.16
$A/in.^2$	A/mm^2	0.00155

SYNONYMS, SYMBOLS, ABBREVIATIONS, AND RELATED TERMS:
Superconductivity, Critical Current Density

Critical Field Strength of Superconductivity

Common Abbreviation:	H_c
Preferred or SI Unit	A/m
Alternate or English Unit	Oe

DEFINITION:

The Critical Field Strength of Superconductivity defines an upper threshold for a stable state of superconductivity. Type I superconductors are superconductive below the critical value H_c, while Type II superconductors have two different critical field values identified as H_{c1} and H_{c2}. Type II superconductors exhibit perfect diamagnetism like Type I superconductors when the field strength of the Type II is below H_{c1}. Type II materials remain superconductive up to a higher value H_{c2}, which is larger than H_c for Type I superconductors.

MATERIAL/TEST PARAMETERS: Current density, temperature

CONVERSION FACTORS:

to convert	to	multiply by
A/m	Oe	0.01256637
Oe	A/m	79.57747

SYNONYMS, SYMBOLS, ABBREVIATIONS, AND RELATED TERMS: Critical Magnetic Field, Critical Field Strength, H_{c1}, H_{c2}

Critical Temperature of Superconductivity

Common Abbreviation:	T_c
Preferred or SI Unit	K
Alternate or English Unit	°F

DEFINITION:

The Critical Temperature of Superconductivity defines one of the three upper thresholds for a stable state of superconductivity. A state of superconductivity is stable only if the temperature, magnetic field strength, and current density are below their critical values.

MATERIAL/TEST PARAMETERS: Current density, magnetic field strength

CONVERSION FACTORS:

to convert	to	multiply by
K	°F	1.8 then subtract 459.67
°F	K	0.55556 then add 225.37
°C	K	1 then subtract 273.15
R	K	0.55556

SYNONYMS, SYMBOLS, ABBREVIATIONS, AND RELATED TERMS: Critical Temperature

Dissipation Factor

Common Abbreviation: D

Preferred or SI Unit (Unitless)

DEFINITION:

The Dissipation Factor (D) is the ratio of the Loss Index to its Relative Permittivity. D is also the reciprocal of the Quality Factor (Q, sometimes called the Storage Factor) of electrical insulators. The Loss Index is the magnitude of the imaginary part of the relative complex permittivity. Relative Permittivity is the real part of the relative complex permittivity. See ASTM D 1711 for other terms having general application relating to electrical insulation.

EXAMPLE TEST METHOD: ASTM D 1711

SYNONYMS, SYMBOLS, ABBREVIATIONS, AND RELATED TERMS: Loss Tangent, Quality Factor Reciprocal, Q^{-1}, tan δ, Quality Factor, Q, Storage Factor, Loss Factor, Loss Index, κ'', Relative Permittivity, Permittivity, Dielectric Constant, SIC, Complex Permittivity

Electrical Conductivity

Common Abbreviation:	σ
Preferred or SI Unit	10^6 S/m
Alternate or English Unit	%IACS

DEFINITION:

Electrical Conductivity is related to the ease of passage of electrical current through matter. Electrical Conductivity is defined as the electrical charge flux per unit voltage gradient, or the reciprocal of Electrical Resistivity expressed in units of (ohm length)$^{-1}$. The conductivity of a metal or alloy is commonly compared with that of the International Annealed Copper Standard (IACS), and its conductivity is then expressed as %IACS. (*Note: Conversion factors below, to, and from %IACS are only valid at 20 °C.) Conductance (expressed as (ohm)$^{-1}$ or S) is the ratio of the current carried through a material to the difference in potential applied across the material. Conductance is the reciprocal of Resistance.

MATERIAL/TEST PARAMETERS: Temperature

CONVERSION FACTORS:

to convert	to	multiply by
10^6 S/m	%IACS	1.724 *See Note above
%IACS	10^6 S/m	0.58001 *See Note above
10^6(mho/m)	10^6 S/m	1

SYNONYMS, SYMBOLS, ABBREVIATIONS, AND RELATED TERMS:
Conductivity, Conductance

EXAMPLE VALUES:	**10^6 S/m**	**%IACS**
Carbon and Alloy Steels	1-10	3-17
Aluminum Alloys	15-38	27-65
Copper Alloys	3-60	5-104
Stainless Steels	0.5-1	2-3

Relative Resistivity

Common Abbreviation: (None established)

Preferred or SI Unit (Unitless)

DEFINITION:

Relative Resistivity is the ratio of resistivity with respect to a reference resistivity (such as the resistivity at room temperature). Relative Resistivity often is used to express resistivity changes with temperature.

MATERIAL/TEST PARAMETERS: Temperature, reference resistivity

Resistivity

Common Abbreviation:	ρ
Preferred or SI Unit	μΩ · m
Alternate or English Unit	Ω circular-mil/ft

DEFINITION:

Resistivity (ρ) is that property of a material which determines its resistance to the flow of an electric current. Resistivity is equal to resistance (R) of the specimen in ohms (Ω) multiplied by the cross-sectional area and divided by the length of the specimen. The value of ρ is equivalent to the resistance between opposite faces of a cube of unit dimensions and can be designated Specific Resistivity or Volume Resistivity. Resistance, expressed in ohms, is the ratio of the potential difference applied to a specimen to the current passed through by the applied potential. Resistance is the reciprocal of Conductance.

MATERIAL/TEST PARAMETERS: Temperature

CONVERSION FACTORS:

to convert	to	multiply by
μΩ · m	Ω circular-mil/ft	601.5305
Ω circular-mil/ft	μΩ · m	0.001662426
μΩ· cm	μΩ · m	0.01
μΩ · in.	μΩ · m	0.0254
$\Omega mm^2/m$	μΩ · m	1

SYNONYMS, SYMBOLS, ABBREVIATIONS, AND RELATED TERMS: Electrical Resistivity, Volume Resistivity, Specific Resistivity, Resistance, R

EXAMPLE VALUES:	**μΩ · m**	**Ω circular-mil/ft**
Carbon and Alloy Steels	0.15-0.23	90-140
Aluminum Alloys	0.03-0.06	18-36
Copper Alloys	0.017-0.28	10-170
Stainless Steels	0.6-1.1	360-660
Titanium Alloys	0.4-1.7	24-1020

Thermal Coefficient of Resistivity

Common Abbreviation:	$\Delta\rho/\rho$
Preferred or SI Unit	$°C^{-1}$
Alternate or English Unit	$°F^{-1}$

DEFINITION:

The Thermal Coefficient of Resistivity is the slope or fractional change in Resistivity per unit change in temperature. If the change in Resistivity is not linear, then the coefficient must be reported as a function of temperature.

MATERIAL/TEST PARAMETERS: Temperature

CONVERSION FACTORS:

to convert	to	multiply by
$°C^{-1}$	$°F^{-1}$	0.55556
$°F^{-1}$	$°C^{-1}$	1.8

SYNONYMS, SYMBOLS, ABBREVIATIONS, AND RELATED TERMS:
Temperature Coefficient of Resistivity, Temperature Coefficient of Resistance

Thermoelectric Potential

Common Abbreviation: (None established)

Preferred or SI Unit mV

DEFINITION:

The Thermoelectric Potential is the voltage due to differences in temperature between two junctions of dissimilar metals.

MATERIAL/TEST PARAMETERS: Metal at junction

SYNONYMS, SYMBOLS, ABBREVIATIONS, AND RELATED TERMS: Seebeck Potential, Seebeck Electromotive Force, Thermal Electromotive Force

Weight Conductivity

Common Abbreviation:	(None established)
Preferred or SI Unit	S m^2/g
Alternate or English Unit	Ω $mile^2$/lb

DEFINITION:

Weight Conductivity refers to the inherent conductance of a material normalized to its density. Weight Conductivity is the reciprocal of Weight Resistivity.

CONVERSION FACTORS:

to convert	to	multiply by
S m^2/g	Ω $mile^2$/lb	0.000175
Ω $mile^2$/lb	S m^2/g	5,710
Ω ft^2/lb	S m^2/g	1.592×10^{11}
%IACS (weight basis)	S m^2/g	(depends on density relative to Cu)

SYNONYMS, SYMBOLS, ABBREVIATIONS, AND RELATED TERMS:
Conductivity (weight basis)

Weight Resistivity

Common Abbreviation: (None established)

Preferred or SI Unit Ω g/m^2
Alternate or English Unit Ω lb/mile^2

DEFINITION:

Weight Resistivity refers to the inherent electrical resistance of a material normalized to its density. Weight Resistivity is the product of Volume Resistivity and Density.

CONVERSION FACTORS:

to convert	to	multiply by
Ω g/m^2	Ω lb/mile^2	5,710
Ω lb/mile^2	Ω g/m^2	0.00017513
Ω lb/ft^2	Ω g/m^2	1.592×10^{11}

SYNONYMS, SYMBOLS, ABBREVIATIONS, AND RELATED TERMS: Electrical Resistivity (weight basis)

Coercive Force

Common Abbreviation:	H_c
Preferred or SI Unit	A/m
Alternate or English Unit	Oe

DEFINITION:

Coercive Force (H_c) is a measure of the magnetic retentivity of magnetic materials. Permanent magnets have relatively high values. H_c is the magnetizing force (Magnetic Field Strength, H) that must be applied in the direction opposite to that of the previous magnetizing force in order to reduce Magnetic Flux Density (induction, B) to zero. Intrinsic Coercive Force (H_{ci}) refers to the field strength needed to reduce Intrinsic Magnetic Induction (B_i) to zero.

EXAMPLE TEST METHODS: ASTM A 341/A 34, ASTM A 596/A 59

CONVERSION FACTORS:

to convert	to	multiply by
A/m	Oe	0.01256637
Oe	A/m	79.57747

SYNONYMS, SYMBOLS, ABBREVIATIONS, AND RELATED TERMS: Magnetic Field Strength, Intrinsic Coercive Force, H_{ci}

EXAMPLE VALUES:	A/m	Oe
Carbon Steel	3,340	42
Permanent Magnet Alloys	300-1,500	24,000-120,000
Cobalt Rare Earth Alloys	7,000 (H_{ci} > 10,000)	557,000 (H_{ci} > 796,000)
Fe-Nd B Alloys	700,000-800,000	9,000-10,000
High-Permeability Fe Alloys	0.01-0.2	0.8-16
Co Magnet Steel	12,000-24,000	150-300

Coercivity

Common Abbreviation:	H_{cs}
Preferred or SI Unit	A/m
Alternate or English Unit	Oe

DEFINITION:

Coercivity, which is the maximum Coercive Force of a material, refers to the Magnetic Field Strength (H) necessary to reduce the Magnetic Flux Density (B) to zero from its saturated value, B_s.

CONVERSION FACTORS:

to convert	to	multiply by
A/m	Oe	0.01256637
Oe	A/m	79.57747

Curie Temperature

Common Abbreviation:	T_C
Preferred or SI Unit	°C
Alternate or English Unit	°F

DEFINITION:

The Curie Temperature is the temperature at which ferromagnetic materials can no longer be magnetized by outside forces, and at which they lose their residual magnetism. The Curie Temperature is also the temperature of magnetic transformation below which a metal or alloy is ferromagnetic, and above which it is paramagnetic.

CONVERSION FACTORS:

to convert	to	multiply by	then add
°C	°F	1.8	32
°F	°C	0.55556	–17.7778
K	°C	1	–273.15
R	°C	0.55556	273.15

SYNONYMS, SYMBOLS, ABBREVIATIONS, AND RELATED TERMS: Curie Point

EXAMPLE VALUES:	**°C**	**°F**
Permanent Magnet Alloys	450-900	850-1,650
Soft Magnetic Iron	300-760	600-1,400
Fe Nd B Alloys	300	600
Ni Fe Alloys	230-500	450-950
Electrical Steels (Fe-Si)	750	1,375

Hall Coefficient

Common Abbreviation:	R_H
Preferred or SI Unit	Vm^2/A^2
Alternate or English Unit	n Ω cm/Oe

DEFINITION:

The Hall Effect is the development of a transverse electric field in a current-carrying conductor placed in a magnetic field. The Hall Coefficient (R_H) is a coefficient in the Hall Effect equation [$V_H = (R_H I B)/d$] that relates the Hall voltage (V_H), the current (I), the magnetic field (B), and the specimen thickness (d). R_H is a constant for a given material and is negative for metals.

CONVERSION FACTORS:

to convert	to	multiply by
Vm^2/A^2	n Ω cm/Oe	7.957747×10^{12}
n Ω cm/Oe	Vm^2/A^2	1.256637×10^{-13}

SYNONYMS, SYMBOLS, ABBREVIATIONS, AND RELATED TERMS: Hall Effect Coefficient, Hall Constant

Intrinsic Magnetic Induction

Common Abbreviation:	B_i
Preferred or SI Unit	T
Alternate or English Unit	G

DEFINITION:

Intrinsic Magnetic Induction (B_i) is the difference between the magnetic induction in a magnetic material and the magnetic induction in a vacuum under the influence of the same magnetizing force. In cgs units, $B_i/4\pi$ is often called Magnetic Polarization (J). Magnetization (M) is the component of the total magnetizing force that produces the Intrinsic Induction in a magnetic material.

CONVERSION FACTORS:

to convert	to	multiply by
T	G	10,000
G	T	0.0001

SYNONYMS, SYMBOLS, ABBREVIATIONS, AND RELATED TERMS: Intrinsic Induction, Magnetic Polarization, Intrinsic Flux Density, Magnetization, M, J, Saturation Induction, B_s

Magnetic Core Loss

Common Abbreviation:	P_c
Preferred or SI Unit	W/kg
Alternate or English Unit	W/lb

DEFINITION:

Magnetic Core Loss is the active power (P_c) expended per unit mass of magnetic material with cyclically varying induction at a specific frequency.

EXAMPLE TEST METHODS: ASTM A 340, ASTM A 596

MATERIAL/TEST PARAMETERS: Magnetic Induction (B), frequency

CONVERSION FACTORS:

to convert	to	multiply by
W/kg	W/lb	0.45359
W/lb	W/kg	2.250

SYNONYMS, SYMBOLS, ABBREVIATIONS, AND RELATED TERMS: Core Loss

EXAMPLE VALUES:	**W/kg**	**W/lb**
Electric Steels (B = 1.5 T)	1-13	0.5-6

Magnetic Energy Product

Common Abbreviation:	BH
Preferred or SI Unit	TA/m
Alternate or English Unit	GOe

DEFINITION:

The Magnetic Energy Product is the product of the values of Magnetic Induction (B) and the Magnetic Field Strength (H) at any point on a demagnetization curve. The Magnetic Energy Product is a measure of the magnetic energy available for use outside the magnet body. The Maximum or Peak Energy Product (BH_{max}) indicates the maximum magnetic energy available and is usually cited as a figure of merit for determining the quality of permanent magnet materials.

CONVERSION FACTORS:

to convert	to	multiply by
TA/m	GOe	125.6637
GOe	TA/m	0.00795775

SYNONYMS, SYMBOLS, ABBREVIATIONS, AND RELATED TERMS: B_dH_d, Energy Product, Peak Energy Product, Maximum Energy Product, BH_{max}

EXAMPLE VALUES:	**TA/m**	**GOe**
Carbon Steel BH_{max}	1,400	0.18×10^6
Permanent Magnet Alloys BH_{max}	8,000-60,000	$1\text{-}7.5 \times 10^6$
Pt Co Alloys BH_{max}	75,600	9.5×10^6
Fe-Nd B Alloys BH_{max}	75,600-279, 000	$9.5\text{-}35 \times 10^6$

Magnetic Field Strength

Common Abbreviation:	H
Preferred or SI Unit	A/m
Alternate or English Unit	Oe

DEFINITION:

Magnetic fields are quantified in terms of their strength, H, which is related to the force produced by a pair of magnets. One oersted (Oe), for example, corresponds to a repulsive force of one dyne acting on a unit test pole by similar test pole at a distance of one centimeter. The SI unit is defined in terms of the magnetic field inside a long solenoid having N turns per meter and carrying an electric current. Demagnetizing Field Strength (H_d) is applied in such a direction as to reduce the induction in a magnetized body. H measures the ability of magnetized bodies to produce magnetic induction at a given point, and a maximum value (H_{max}, Peak Magnetizing Force, or H_p) is frequently reported for permanent magnet alloys.

EXAMPLE TEST METHODS: ASTM A 341, ASTM A 348, ASTM A 596

CONVERSION FACTORS:

to convert	to	multiply by
A/m	Oe	0.01256637
Oe	A/m	79.57747

SYNONYMS, SYMBOLS, ABBREVIATIONS, AND RELATED TERMS: Magnetic Field Intensity, Magnetic Force, Magnetic Intensity, Magnetizing Force, Demagnetizing Force, Demagnetizing Field Strength, H_d, Peak Magnetizing Force, H_{max}, H_p

EXAMPLE VALUES:	**A/m**	**Oe**
Permanent Magnet Alloys H_{max}	79,600-318,400	1,000-4,000

Magnetic Field Strength at BH_{max}

Common Abbreviation:	(None established)
Preferred or SI Unit	A/m
Alternate or English Unit	Oe

DEFINITION:

Magnetic Field Strength at BH_{max} is the Magnetic Field Strength (H) of the point on the B versus H demagnetization curve that, when multiplied by the corresponding coordinate of Magnetic Flux Density (B), gives the maximum value of Magnetic Energy Product (BH) for the curve. The most efficient design for a magnet is one that employs a field strength of the BH_{max} value.

CONVERSION FACTORS:

to convert	to	multiply by
A/m	Oe	0.01256637
Oe	A/m	79.57747

SYNONYMS, SYMBOLS, ABBREVIATIONS, AND RELATED TERMS:
Demagnetizing Field at BH_{max}, Magnetizing Force at BH_{max}

EXAMPLE VALUES:	**A/m**	**Oe**
Permanent Magnet Alloys	8,000-40,000	100-500

Magnetic Hysteresis Loss

Common Abbreviation:	(None established)
Preferred or SI Unit	J/m^3 per cycle
Alternate or English Unit	erg/cm^3 per cycle

DEFINITION:

Magnetic Hysteresis Loss is the energy loss per cycle in a magnetic material as a result of magnetic hysteresis.

CONVERSION FACTORS:

to convert	to	multiply by
J/m^3 per cycle	erg/cm^3 per cycle	10^{13}
erg/cm^3 per cycle	J/m^3 per cycle	10^{-13}

SYNONYMS, SYMBOLS, ABBREVIATIONS, AND RELATED TERMS: Hysteresis Loss

EXAMPLE VALUES:	**J/m^3 per cycle**	**erg/cm^3 per cycle**
Magnetically Soft Ni Fe Alloys	$9\text{-}20 \times 10^{-13}$	9-20

Magnetic Induction

Common Abbreviation:	B
Preferred or SI Unit	T
Alternate or English Unit	G

DEFINITION:

Magnetic Induction (B) is often called Flux Density, and is the strength of the induced field that permeates the space near a magnet. When a material is placed in a magnetic field, $B = \mu_0 \mu_r H$ where H is the Magnetic Field Strength, μ_0 is the Magnetic Permeability of a vacuum, and μ_r is the Relative Permeability. B is a vector quantity measured either by the mechanical force experienced by an element of electric current, or by the electromotive force induced in an elementary loop during any change in flux linkages with the loop.

EXAMPLE TEST METHODS: ASTM A 341, ASTM A 596

CONVERSION FACTORS:

to convert	to	multiply by
T	G	10,000
G	T	0.0001
Wb/m^2	T	1
$V \cdot s/m^2$	T	1

SYNONYMS, SYMBOLS, ABBREVIATIONS, AND RELATED TERMS: Magnetic Flux Density, Flux Density, Magnetic Displacement, Magnetic Vector, Induction

Magnetic Induction at BH_{max}

Common Abbreviation:	B_0
Preferred or SI Unit	T
Alternate or English Unit	G

DEFINITION:

Magnetic Induction at BH_{max} is the Magnetic Flux Density coordinate (B) of the point on the B versus H demagnetization curve that, when multiplied by the corresponding coordinate of Magnetic Field Strength (H), gives the maximum value of Magnetic Energy Product (BH) for the curve. Magnetic Induction at BH_{max} is also the value of B at BH_{max} on a plot of B versus BH. The most efficient design for a magnet is one that employs magnetic induction corresponding to the BH_{max} value.

MATERIAL/TEST PARAMETERS: B/H ratio

CONVERSION FACTORS:

to convert	to	multiply by
T	G	10,000
G	T	0.0001
Wb/m^2	T	1
$V \cdot s/m^2$	T	1

SYNONYMS, SYMBOLS, ABBREVIATIONS, AND RELATED TERMS: Magnetic Flux Density at BH_{max}, Flux Density at BH_{max}, Induction at BH_{max}

EXAMPLE VALUES:	**T**	**G**
Permanent Magnet Alloys	0.3-1.2	3,000-12,000

Magnetic Peak Inductance

Common Abbreviation:	B_{max}
Preferred or SI Unit	T
Alternate or English Unit	G

DEFINITION:

Magnetic Peak Inductance is the value of Magnetic Induction (B) on a magnetic material's hysteresis loop at which B no longer increases with increasing Magnetic Field Strength (H). Magnetic Peak Inductance is also the Maximum Intrinsic Induction possible in a material.

CONVERSION FACTORS:

to convert	to	multiply by
T	G	10,000
G	T	0.0001
Wb/m^2	T	1
$V \cdot s/m^2$	T	1

SYNONYMS, SYMBOLS, ABBREVIATIONS, AND RELATED TERMS: Saturation Induction, Maximum Intrinsic Induction, Peak Inductance, Saturation Flux Density, Magnetic Saturation

EXAMPLE VALUES:	T	G
Permanent Magnet Alloys	10-16	10,000-16,000
Soft Magnetic Alloys	7-9	7,000-9,000
Electrical Steel	1.8	1,800

Magnetic Permeability

Common Abbreviation:	μ
Preferred or SI Unit	Unitless (cgs)
Alternate or English Unit	G/Oe

DEFINITION:

Magnetic Permeability (μ) is a generic term used to express relationships ($B = \mu H$) between Magnetic Induction (B) and Magnetic Field Strength (H). Permeability is a quantity that indicates the ease of inducing magnetic flux lines in a material. The quantity μ is typically expressed as $\mu = \mu_0 \mu_r$, where μ_0 is the permeability of a vacuum and μ_r is the Relative Permeability of a substance. In cgs units, the relation is simplified to $\mu = \mu_r$, because $\mu_0 = 1$ when B and H are given in the cgs units of gauss and oersteds. For any ferromagnetic material, Permeability is a function of the degree of magnetization, but Intrinsic Permeability (μ_0) and Maximum Permeability (μ_m) are unique values for a given specimen under specific conditions. Intrinsic Permeability (μ_i) is the ratio of Intrinsic Magnetic Induction (B_i) to the corresponding magnetic Field Strength. Absolute Permeability (μ_{abs}) is the ratio of the total induction (ΔB) to the field strength (ΔH) which produces it. Incremental Permeability (μ_Δ) is the ratio of change in Magnetic Induction to the corresponding change in Magnetic Field Strength under certain conditions. See also Relative Permeability.

EXAMPLE TEST METHODS: ASTM A 341, ASTM A 346, ASTM A 596

MATERIAL/TEST PARAMETERS: Hysteresis (previous history), Magnetic Field Strength, test conditions

SYNONYMS, SYMBOLS, ABBREVIATIONS, AND RELATED TERMS: Permeability, μ_0, μ_m, Initial Permeability, Maximum Permeability, Intrinsic Permeability, μ_i, Absolute Permeability, μ_{abs}, B/H, μ_Δ, Incremental Permeability

EXAMPLE VALUES:	**Unitless (cgs)**
Soft Magnetic Ni Fe Alloys	μ_m 80,000-400,000
Ferritic Stainless Steel	μ_m 1,000-3,000
Martensitic Stainless Steel	μ_m 50-950
Permanent Magnet Alloys	μ_Δ 1-8

Magnetic Residual Induction

Common Abbreviation:	B_r
Preferred or SI Unit	T
Alternate or English Unit	G

DEFINITION:

Magnetic Residual Induction (B_r) is the magnetic induction corresponding to a zero magnetizing field when a material is subjected to certain magnetized conditions. A permanent magnet has a relatively high value for B_r. Remanent Induction (B_d) is the magnetic induction that remains in a magnetic circuit after the removal of an applied magnetic field. Remanence (B_{dm}) is the maximum value of Remanent Induction for a given geometry of a magnetic circuit. Retentivity (B_{rs}) is usually associated with Saturation Induction and is the maximum value of Residual Induction. $B_r = B_d$, and $B_{rs} - B_{dm}$, if there are no inhomogeneities in the magnetic circuit; if air gaps or other inhomogeneities are present then $B_d < B_r$, and $B_{dm} < B_{rs}$.

MATERIAL/TEST PARAMETERS: Circuit geometry, test conditions

CONVERSION FACTORS:

to convert	to	multiply by
T	G	10,000
G	T	0.0001
Wb/m^2	T	1
$V \cdot s/m^2$	T	1

SYNONYMS, SYMBOLS, ABBREVIATIONS, AND RELATED TERMS: Residual Induction, Remanence, Retentivity, Residual Magnetic Inductance, Remanent Magnetization, Magnetic Remanence, Residual Flux Density, B_d, B_{dm}, B_{rs}, B_d, Remanent Induction, Residual Magnetism

EXAMPLE VALUES:	**T**	**G**
Permanent Magnet Alloys	0.5-1.4	5,000-14,000
Silver Alloys	0.059	590
High Permeability Iron	3-5	3,000-5,000

Magnetic Susceptibility

Common Abbreviation: κ

Preferred or SI Unit (Unitless)

DEFINITION:

Magnetic Susceptibility is defined as the ratio of magnetization (Intrinsic Magnetic Induction) to the Magnetic Field Strength (B_i/H or M/H), which is also equivalent to the Relative Permeability minus one. If values are slightly positive, the material is paramagnetic; if slightly negative, diamagnetic. Ferromagnetic materials have Magnetic Susceptibilities ranging from 10^2 to 10^6. Mass Susceptibility is the preferred form of reported data. See Mass Susceptibility and also Molar Susceptibility and Volume Susceptibility.

Magnetizing Field Requirement

Common Abbreviation:	(None established)
Preferred or SI Unit	A/m
Alternate or English Unit	Oe

DEFINITION:

The Magnetizing Field Requirement is the field strength required to magnetize a virgin magnet (a magnet that has never been exposed to a magnetizing field after the final heat treatment or other processing/fabricating operations). In general, as a magnetic material's Coercive Force increases, the required magnetizing field also increases, and, in some cases, even exceeds the Magnetic Peak Inductance.

CONVERSION FACTORS:

to convert	to	multiply by
A/m	Oe	0.01256637
Oe	A/m	79.57747

SYNONYMS, SYMBOLS, ABBREVIATIONS, AND RELATED TERMS: Required Magnetizing Field

Magnetoresistance Coefficient

Common Abbreviation:	(None established)
Preferred or SI Unit	m^2/A^2
Alternate or English Unit	Oe^{-2}

DEFINITION:

The Magnetoresistance Coefficient is the fractional change in Resistivity ($\Delta\rho/\rho$) when a magnetic field (H) is applied. The change in resistance usually is proportional to H^2 unless the field strength is very large. For strong magnetic fields, the Magnetoresistance Coefficient may be proportional to H instead of H^2.

CONVERSION FACTORS:

to convert	to	multiply by
m^2/A^2	Oe^{-2}	6332.574
Oe^{-2}	m^2/A^2	0.0001579

SYNONYMS, SYMBOLS, ABBREVIATIONS, AND RELATED TERMS:
Magnetoresistivity Coefficient

Mass Susceptibility

Common Abbreviation:	χ
Preferred or SI Unit	cm^3/g

DEFINITION:

Mass Susceptibility is the Magnetic Susceptibility divided by Density and is the preferred form of reported data for Magnetic Susceptibility.

Molar Susceptibility

Common Abbreviation: χ_M

Preferred or SI Unit $cm^3/mole$

DEFINITION:

Molar Susceptibility is the Magnetic Susceptibility divided by moles per cubic centimeter.

SYNONYMS, SYMBOLS, ABBREVIATIONS, AND RELATED TERMS: Atomic Susceptibility, χ_A

Relative Permeability

Common Abbreviation: μ_r

Preferred or SI Unit (Unitless)

DEFINITION:

The Relative Permeability (μ_r) of a material is the ratio of Absolute Permeability to the magnetic constant. The magnetic constant (Γ_m) is the dimensionless scalar factor that relates the mechanical force between two currents to their intensities and geometric configuration. Relative Permeability is numerically the same as Absolute Permeability if cgs units are used. See also Magnetic Permeability.

EXAMPLE TEST METHOD: ASTM A 342

MATERIAL/TEST PARAMETERS: Field strength (H)

Volume Susceptibility

Common Abbreviation:	χ_V
Preferred or SI Unit	cm^3

DEFINITION:

Volume Susceptibility is the Magnetic Susceptibility divided by volume.

Compressibility

Common Abbreviation:	κ
Preferred or SI Unit	Pa^{-1}
Alternate or English Unit	$in.^2/lbf$

DEFINITION:

Compressibility (κ) relates the fractional change in volume (ΔV/V) of a substance with a given increase in pressure (P): $\kappa = \Delta V/V \times \Delta P$. Compressibility is the reciprocal of Bulk Modulus in the elastic range.

EXAMPLE TEST METHOD: ASTM B 331 (metal powders)

CONVERSION FACTORS:

to convert	to	multiply by
Pa^{-1}	$in.^2/lbf$	6894.757
$in.^2/lbf$	Pa^{-1}	0.000145
cm^2/dyn	Pa^{-1}	10
m^2/N	Pa^{-1}	1

SYNONYMS, SYMBOLS, ABBREVIATIONS, AND RELATED TERMS:
Compactibility

EXAMPLE VALUES:	**Pa^{-1}**	**$in.^2/lbf$**
Iron	5.91×10^{-12}	40.7×10^{-9}
Aluminum	13.3×10^{-12}	91.7×10^{-9}
Copper	7.25×10^{-12}	50×10^{-9}
Magnesium	28.1×10^{-12}	194×10^{-9}
Titanium	9.22×10^{-12}	63.6×10^{-9}

Density

Common Abbreviation:	ρ
Preferred or SI Unit	kg/m^3
Alternate or English Unit	$lb/in.^3$

DEFINITION:

Density is the mass per unit volume of a material at a specified temperature.

MATERIAL/TEST PARAMETERS: Temperature

CONVERSION FACTORS:

to convert	to	multiply by
kg/m^3	$lb/in.^3$	0.000036
$lb/in.^3$	kg/m^3	27,679.9
Mg/m^3	kg/m^3	1,000
lb/ft^3	kg/m^3	16.02
g/cm^3	kg/m^3	1,000

SYNONYMS, SYMBOLS, ABBREVIATIONS, AND RELATED TERMS: Mass per Unit Volume

EXAMPLE VALUES:	kg/m^3	$lb/in.^3$
Carbon and Alloy Steels	7,850	0.283
Aluminum Alloys	2,710	0.0978
Copper Alloys	8,940	0.3227
Magnesium Alloys	1,810	0.0653
Stainless Steels	7,900	0.2852
Titanium Alloys	4,430	0.1660

Dynamic Viscosity

Common Abbreviation:	η
Preferred or SI Unit	Pa · s
Alternate or English Unit	cP

DEFINITION:

Dynamic Viscosity is the force required to overcome fluid friction. Dynamic Viscosity is defined as the ratio between applied shear stress and the rate of shear.

CONVERSION FACTORS:

to convert	to	multiply by
Pa · s	cP	1,000
cP	Pa · s	0.001
poise (P)	Pa · s	0.1
lb · mass/ft · s	Pa · s	1.488
lbf · s/ft^2	Pa · s	47.88

SYNONYMS, SYMBOLS, ABBREVIATIONS, AND RELATED TERMS:
Coefficient of Dynamic Viscosity, Absolute Viscosity

EXAMPLE VALUES:	**Pa · s**	**cP**
Carbon and Alloy Steels	6.9×10^{-3}	6.9
Aluminum Alloys	1.2×10^{-3}	1.2
Copper Alloys	4.3×10^{-3}	4.3
Magnesium Alloys	1.25×10^{-3}	1.25

Kinematic Viscosity

Common Abbreviation:	ν
Preferred or SI Unit	m^2/s
Alternate or English Unit	cSt (centistoke)

DEFINITION:

Kinematic Viscosity (ν) is related to Dynamic Viscosity (η) by the equation: $\nu = \eta/\rho$, where ρ is the density. Kinematic Viscosity can also be thought of as momentum diffusivity, analogous to thermal diffusivity and mass diffusivity.

EXAMPLE TEST METHOD: ASTM D 2864

MATERIAL/TEST PARAMETERS: Temperature

CONVERSION FACTORS:

to convert	to	multiply by
m^2/s	cSt	10^6
cSt	m^2/s	10^{-6}
ft^2/s	m^2/s	0.0929

EXAMPLE VALUES:	**m^2/s**	**cSt (centistoke)**
Iron (at the melting point)	0.982×10^{-6}	0.982
Aluminum (at the melting point)	0.504×10^{-6}	0.504
Copper (at the melting point)	0.563×10^{-6}	0.563
Magnesium (at the melting point)	0.786×10^{-6}	0.786

Velocity of Sound

Common Abbreviation: c

Preferred or SI Unit: km/s
Alternate or English Unit: ft/s

DEFINITION:

Velocity of Sound is the speed of mechanical vibrations traveling through a substance. The Velocity of Sound (c) is related to density (ρ) and the Bulk Modulus (K) as follows: $c = \sqrt{K}/\rho$. The Velocity of Sound depends on whether the waveform consists of longitudinal, transverse, or shear waves, which have speeds indicated as c_L, c_T, and c_S respectively. The group velocity (c_g) is related to the dispersion in multiple-frequency waveforms.

MATERIAL/TEST PARAMETERS: Waveform

CONVERSION FACTORS:

to convert	to	multiply by
km/s	ft/s	3,280.84
ft/s	km/s	0.0003048
m/s	km/s	0.001

SYNONYMS, SYMBOLS, ABBREVIATIONS, AND RELATED TERMS: Sonic Velocity, Speed of Sound

EXAMPLE VALUES:	km/s	ft/s
Carbon and Alloy Steels	Longitudinal waves in bulk material: 6.0	19,600
Aluminum Alloys	6.4	21,000
Copper Alloys	5.0	16,400
Magnesium Alloys	5.8	19,000
Stainless Steels	6.0	19,600
Titanium Alloys	6.0	19,600

Grain Size

Common Abbreviation: (None established)

DEFINITION:

Grain Size is an estimation of the dimensions of the grains or crystals in a metal. Hardness and strength of polycrystalline metals usually have an inverse relationship with Grain Size. Numerous methods, with results expressed in different units, have been developed to provide Grain Size measurements.

EXAMPLE TEST METHODS: ASTM E 7, ASTM E 1112, ASTM A 919

MATERIAL/TEST PARAMETERS: Units, test method

SYNONYMS, SYMBOLS, ABBREVIATIONS, AND RELATED TERMS: G, ASTM Grain Size Number (G), Ferritic Grain Size, Austenitic Grain Size, Fracture Grain Size, Grain Diameter, $\bar{d}$, Grain Area, $\bar{A}$, Grains per Unit Area, N_A, Grains per Unit Volume, N_V

Lattice Parameter

Common Abbreviation:	(None established)
Preferred or SI Unit	for Unit Cell Length: nm
Alternate or English Unit	for Unit Cell Length: angstrom

DEFINITION:

The term Lattice Parameter is used for the fractional coordinates x, y, and z of lattice points when these are variable, or to indicate the lengths of the axes a, b, and c, and their included angles α, β, and γ.

CONVERSION FACTORS:

to convert	to	multiply by
nm	angstrom	10
angstrom	nm	0.1

SYNONYMS, SYMBOLS, ABBREVIATIONS, AND RELATED TERMS: Crystal Lattice Length, Lattice Constant, Unit Cell Length, Unit Cell Angle

Optical Reflectance

Common Abbreviation:	ρ
Preferred or SI Unit	%
Alternate or English Unit	Unitless

DEFINITION:

Optical Reflectance is defined as the ratio of the intensity of reflected light to the intensity of incoming light. This property is a function of the wavelength of the incoming radiation.

MATERIAL/TEST PARAMETERS: Wavelength

CONVERSION FACTORS:

to convert	to	multiply by
%	unitless	0.01
unitless	%	100

SYNONYMS, SYMBOLS, ABBREVIATIONS, AND RELATED TERMS:
Reflectance, Coefficient of Reflection, Reflectivity, Reflection Factor

Surface Energy

Common Abbreviation:	(None established)
Preferred or SI Unit	J/m^2
Alternate or English Unit	erg/cm^2

DEFINITION:

Surface Energy is the work required to increase the surface area by 1 cm^2. Because surface atoms are attracted toward the center of a liquid or solid, in order to increase the surface, work must be done to overcome the inward pull. Surface Energy is a function of temperature, purity, and crystal orientation (solids). See also Surface Tension.

MATERIAL/TEST PARAMETERS: Temperature

CONVERSION FACTORS:

to convert	to	multiply by
J/m^2	erg/cm^2	1,000
erg/cm^2	J/m^2	0.001
N/m	J/m^2	1

SYNONYMS, SYMBOLS, ABBREVIATIONS, AND RELATED TERMS: σ, γ

EXAMPLE VALUES:	**J/m^2**	**erg/cm^2**
Liquid Iron	1.79	1,790
Liquid Aluminum	0.914	914
Liquid Copper	1.28	1,280
Liquid Magnesium	0.559	559
Liquid Titanium	1.65	1,650

Surface Tension

Common Abbreviation:	(None established)
Preferred or SI Unit	N/mm
Alternate or English Unit	dyne/cm

DEFINITION:

Surface Tension is expressed as work in newtons per millimeter. Surface Tension arises from the molecular forces of the surface film of all liquids that tend to alter the contained volume of liquid into a form of minimum superficial area. See also Surface Energy.

CONVERSION FACTORS:

to convert	to	multiply by
N/mm	dyne/cm	10,000
dyne/cm	N/mm	0.0001

SYNONYMS, SYMBOLS, ABBREVIATIONS, AND RELATED TERMS: σ, γ

Work Function

Common Abbreviation:	Φ
Preferred or SI Unit	eV

DEFINITION:

The Work Function refers to the energy necessary to remove an electron from the surface of a material. Work Function is affected by oxidation and the surface condition.

EXAMPLE TEST METHOD: ASTM E 673

EXAMPLE VALUES:	**eV**
Iron	4.70
Aluminum	4.25
Copper	4.40
Magnesium	3.64
Titanium	3.95

Coefficient of Linear Thermal Expansion

Common Abbreviation:	α_l
Preferred or SI Unit	10^{-6}/K
Alternate or English Unit	10^{-6}/°F

DEFINITION:

The Coefficient of Linear Thermal Expansion (α_l) is the fractional change ($\Delta L/L$) in length (L) accompanying a unit change of temperature, or the tangent at a point on the curve for Linear Thermal Expansion [$\alpha_l = 1/L(dL/dT)$]. The length (L_T) at temperature (T) is related to the length at a reference temperature (L_{ref}) by the equation: $L_T = L_{ref}$ (1 + $(\alpha_l)(\Delta T)$. α_l is a function of temperature and also depends on the specimen orientation when anisotropic crystals or strongly textured materials are being tested. The temperature or temperature range should always be specified when an average value for α_l is reported for that range.

EXAMPLE TEST METHODS: ASTM E 228, ASTM E 289, ASTM E 831

MATERIAL/TEST PARAMETERS: Temperature, orientation

CONVERSION FACTORS:

to convert	to	multiply by
10^{-6}/K	10^{-6}/°F	0.55556
10^{-6}/°F	10^{-6}/K	1.8
ppm/°C	10^{-6}/K	1
10^{-6}/°C	10^{-6}/K	1
(μm/m)/°F	10^{-6}/K	1.8
(μm/m)/°C	10^{-6}/K	1
10^{-6}/R	10^{-6}/K	1.8

SYNONYMS, SYMBOLS, ABBREVIATIONS, AND RELATED TERMS:
Coefficient of Thermal Expansion, Coefficient of Linear Expansion, Thermal Expansion Coefficient, Thermal Coefficient of Expansion, Linear Expansion Coefficient, Linear Thermal Expansion Coefficient, $\overline{\alpha}$, CTE

EXAMPLE VALUES:	**10^{-6}/K**	**10^{-6}/°F**
Carbon and Alloy Steels	11.6	6.45
Aluminum Alloys	23	12.8
Magnesium Alloys	26	14.4

Coefficient of Volumetric Thermal Expansion

Common Abbreviation:	α_v
Preferred or SI Unit	10^{-6}/K
Alternate or English Unit	10^{-6}/°F

DEFINITION:

The Coefficient of Volumetric Thermal Expansion (α_v) is the fractional change of volume accompanying a unit change of temperature. α_v can be estimated as equivalent to $3 \times \alpha_l$, where α_l is the Coefficient of Linear Thermal Expansion of an isotropic material.

MATERIAL/TEST PARAMETERS: Temperature

CONVERSION FACTORS:

to convert	to	multiply by
10^{-6}/K	10^{-6}/°F	0.55556
10^{-6}/°F	10^{-6}/K	1.8
ppm/°C	10^{-6}/K	1
10^{-6}/°C	10^{-6}/K	1
(μm/m)/°F	10^{-6}/K	1.8
(μm/m)/°C	10^{-6}/K	1
10^{-6}/R	10^{-6}/K	1.8

SYNONYMS, SYMBOLS, ABBREVIATIONS, AND RELATED TERMS:
Volumetric Expansion, Thermal Volumetric Expansion, Coefficient of Volumetric Expansion, Thermal Coefficient of Expansion

EXAMPLE VALUES:	**10^{-6}/K**	**10^{-6}/°F**
Carbon and Alloy Steels	34.8	19.3
Aluminum Alloys	69	38.3

Diffusion Coefficient

Common Abbreviation:	D
Preferred or SI Unit	m^2/s
Alternate or English Unit	$in.^2/s$

DEFINITION:

The Diffusion Coefficient is a factor of proportionality (D) that relates the flux of particles (J) in the opposite direction of a concentration gradient (dn/dx) such that: J(x,t) = –D. The flux (J) can be defined as the number of particles (n) crossing, on average, a unit area per second in the direction of increasing x. Diffusion data is identified as self-diffusion, or it is data for a specific element or compound in a given material.

MATERIAL/TEST PARAMETERS: Diffusing element or species

CONVERSION FACTORS:

to convert	to	multiply by
m^2/s	$in.^2/s$	1,550
$in.^2/s$	m^2/s	0.000645
cm^2/s	m^2/s	0.0001

SYNONYMS, SYMBOLS, ABBREVIATIONS, AND RELATED TERMS: Self-diffusion

Diffusion Frequency Factor

Common Abbreviation:	D_0
Preferred or SI Unit	m^2/s
Alternate or English Unit	$in.^2/s$

DEFINITION:

The Diffusion Frequency Factor (D_0) is a coefficient used to model the Diffusion Coefficient (D) by an Arrhenius model such that: $D = D_0 \exp(-Q/RT)$, where Q is the activation energy, R is the molar gas constant, and T is the absolute temperature.

MATERIAL/TEST PARAMETERS: Diffusing element or species

CONVERSION FACTORS:

to convert	to	multiply by
m^2/s	$in.^2/s$	1,550
$in.^2/s$	m^2/s	0.000645
cm^2/s	m^2/s	0.0001

SYNONYMS, SYMBOLS, ABBREVIATIONS, AND RELATED TERMS: Frequency Factor, Pre-exponential Constant, Pre-exponential Frequency Factor, Diffusion Coefficient

Electronic Heat Capacity

Common Abbreviation:	(None established)
Preferred or SI Unit	J/mol · K
Alternate or English Unit	cal/mol · K

DEFINITION:

Electronic Heat Capacity refers to the contribution to Heat Capacity from the motion of conduction electrons. The Electronic Heat Capacity is a significant fraction of the total Heat Capacity only at temperatures very close to absolute zero.

MATERIAL/TEST PARAMETERS: Temperature

CONVERSION FACTORS:

to convert	to	multiply by
J/mol · K	cal/mol · K	0.239
cal/mol · K	J/mol · K	4.184
cal/mol · °C	J/mol · K	4.184

Emittance

Common Abbreviation:	ε
Preferred or SI Unit	Unitless

DEFINITION:

Emittance is the ratio of radiation emitted by a surface to the radiation emitted by a perfect blackbody radiator at the same temperature. Emittance is a term appropriate for a practical body while the term Emissivity is reserved for the special case of an opaque material having a highly polished surface.

EXAMPLE TEST METHODS: ASTM E 408, ASTM E 639

MATERIAL/TEST PARAMETERS: Temperature, wavelength

SYNONYMS, SYMBOLS, ABBREVIATIONS, AND RELATED TERMS:
Emissivity, Thermal Emissivity

EXAMPLE VALUES:	**Unitless (room temperature, wavelength 10^{-6} m)**
Iron	0.41
Aluminum	0.08-0.27
Copper	0.03
Magnesium	0.26
Titanium	0.37-0.49

Enthalpy of Combustion

Common Abbreviation:	(None established)
Preferred or SI Unit	J/mol
Alternate or English Unit	Btu/mol

DEFINITION:

The Enthalpy of Combustion is the amount of heat released in the oxidation of one mole of a substance.

CONVERSION FACTORS:

to convert	to	multiply by
J/mol	Btu/mol	0.0009485
Btu/mol	J/mol	1054.35
cal/mol	J/mol	1054.35

SYNONYMS, SYMBOLS, ABBREVIATIONS, AND RELATED TERMS: Heat of Combustion

EXAMPLE VALUES:	**J/mol**	**Btu/mol**
½(Fe_2O_3)	–412 124	–390.9
½(Al_2O_3)	–837 896	–794.7
CuO	–156 063	–148
MgO	–601 241	–570.2
TiO_2	–944 747	–896

Heat Capacity

Common Abbreviation:	C
Preferred or SI Unit	J/mol · K
Alternate or English Unit	cal/mol · K

DEFINITION:

Heat Capacity (C) indicates a material's ability to absorb heat from the external surroundings. C represents the amount of energy required to produce a unit temperature rise. Heat Capacity is usually expressed per mole of material. See also Specific Heat. C is given the subscript v if the specimen volume is held constant, and the subscript p if the external pressure is held constant.

EXAMPLE TEST METHODS: ASTM D 2766, ASTM E 1269

MATERIAL/TEST PARAMETERS: Temperature, pressure, volume

CONVERSION FACTORS:

to convert	to	multiply by
J/mol · K	cal/mol · K	0.239
cal/mol · K	J/mol · K	4.184
cal/mol · °C	J/mol · K	4.184

SYNONYMS, SYMBOLS, ABBREVIATIONS, AND RELATED TERMS: Thermal Capacity, C_p, C_v, Energy Equivalent

EXAMPLE VALUES:	J/mol · K	cal/mol · K
Iron	25.0	5.98
Aluminum	24.3	5.8
Copper	24.5	5.86
Magnesium	24.7	5.90
Titanium	25.1	5.99

Latent Heat of Evaporation

Common Abbreviation:	(None established)
Preferred or SI Unit	kJ/kg
Alternate or English Unit	Btu/lb

DEFINITION:

Latent Heat is the amount of energy released or absorbed when a substance changes state (such as fusion, vaporization, or other phase transformation).

CONVERSION FACTORS:

to convert	to	multiply by
kJ/kg	Btu/lb	0.4302176
Btu/lb	kJ/kg	2.324405

EXAMPLE VALUES:	kJ/kg	Btu/lb
Iron	6,260	2,692
Aluminum	10,896	4,685
Copper	4,731.6	2,034.6
Magnesium	5,242.8	2,254.4
Titanium	8,562.7	3,682

Latent Heat of Fusion

Common Abbreviation:	(None established)
Preferred or SI Unit	kJ/kg
Alternate or English Unit	Btu/lb

DEFINITION:

Latent Heat is the amount of energy released or absorbed when a substance changes state (such as fusion, vaporization, or other phase transformation).

CONVERSION FACTORS:

to convert	to	multiply by
kJ/kg	Btu/lb	0.4302176
Btu/lb	kJ/kg	2.324405

EXAMPLE VALUES:	**kJ/kg**	**Btu/lb**
Iron	247.2	106.3
Aluminum	397.0	170.7
Copper	206.7	88.9
Magnesium	368.5	158.5
Titanium	277.7	119.4

Liquidus Temperature

Common Abbreviation:	(None established)
Preferred or SI Unit	°C
Alternate or English Unit	°F

DEFINITION:

The Liquidus is the temperature at which a noneutectic alloy begins to freeze on cooling or finish melting on heating. The Liquidus is the upper temperature of a material's Melting Range. Unless otherwise stated, the Liquidus is reported at standard pressure (1 atm).

EXAMPLE TEST METHOD: ASTM E 1142

MATERIAL/TEST PARAMETERS: Pressure

CONVERSION FACTORS:

to convert	to	multiply by	then add
°C	°F	1.8	32
°F	°C	0.55556	–17.7778
K	°C	1	–273.15
R	°C	0.55556	273.15

Melting Point

Common Abbreviation:	T_{mp}
Preferred or SI Unit	°C
Alternate or English Unit	°F

DEFINITION:

The Melting Point is the temperature at which the solid and liquid phases are in equilibrium for a pure substance, compound, or eutectic alloy. When an alloy is noneutectic, melting does not occur at a specific point, but instead occurs over a Melting Range.

CONVERSION FACTORS:

to convert	to	multiply by	then add
°C	°F	1.8	32
°F	°C	0.55556	–17.7778
K	°C	1	–273.15
R	°C	0.55556	273.15

EXAMPLE VALUES:	**°C**	**°F**
Iron	1,537	2,798
Aluminum	660	1,220
Copper	1,083	1,981
Magnesium	650	1,202
Titanium	1,668	3,035

Melting Range

Common Abbreviation:	(None established)
Preferred or SI Unit	°C – °C
Alternate or English Unit	°F – °F

DEFINITION:

The Melting Range is the range of temperatures over which a noneutectic alloy changes from solid to liquid. The Melting Range is the difference between the Solidus Temperature and Liquidus Temperature. Noneutectic alloys can be characterized in terms of a Melting Range, while pure substances and eutectic alloys are characterized by Melting Points.

CONVERSION FACTORS:

to convert	to	multiply by	then add
°C	°F	1.8	32
°F	°C	0.55556	–17.7778
K	°C	1	–273.15
R	°C	0.55556	273.15

Normal Boiling Point

Common Abbreviation:	T_b
Preferred or SI Unit	°C
Alternate or English Unit	°F

DEFINITION:

Normal Boiling Point is the temperature at which the vapor pressure of a liquid equals 10,1325 Pa (1atm).

CONVERSION FACTORS:

to convert	to	multiply by	then add
°C	°F	1.8	32
°F	°C	0.55556	–17.7778
K	°C	1	–273.15
R	°C	0.55556	273.15

SYNONYMS, SYMBOLS, ABBREVIATIONS, AND RELATED TERMS: Boiling Temperature, Boiling Point, T_{VAP}

EXAMPLE VALUES:	**°C**	**°F**
Iron	3,000	5,430
Aluminum	2,450	4,442
Copper	2,595	4,703
Magnesium	1,107	2,025
Titanium	3,260	5,900

Service Temperature (Continuous)

Common Abbreviation:	(None established)
Preferred or SI Unit	°C
Alternate or English Unit	°F

DEFINITION:

The Service Temperature (Continuous) is the maximum temperature to which a material can be continuously exposed without seriously degrading its strength, oxidation resistance, or corrosion resistance in the operating environment. In general, Service Temperature (Continuous) is lower than the Service Temperature (Maximum) and higher than the intermittent service temperature.

MATERIAL/TEST PARAMETERS: Operating environment

CONVERSION FACTORS:

to convert	to	multiply by	then add
°C	°F	1.8	32
°F	°C	0.55556	–17.7778
K	°C	1	–273.15
R	°C	0.55556	273.15

SYNONYMS, SYMBOLS, ABBREVIATIONS, AND RELATED TERMS:
Continuous Service Temperature

Service Temperature (Maximum)

Common Abbreviation:	(None established)
Preferred or SI Unit	°C
Alternate or English Unit	°F

DEFINITION:

The Service Temperature (Maximum) is the maximum temperature to which a material can be exposed without seriously degrading its strength, oxidation resistance, or corrosion resistance in the operating environment. For example, allowable metal temperatures for wrought heat resistant alloys in structural applications generally do not exceed about 950 °C. In non-load bearing applications, allowable temperatures may exceed 1200 °C.

MATERIAL/TEST PARAMETERS: Operating environment

CONVERSION FACTORS:

to convert	to	multiply by	then add
°C	°F	1.8	32
°F	°C	0.55556	–17.7778
K	°C	1	–273.15
R	°C	0.55556	273.15

SYNONYMS, SYMBOLS, ABBREVIATIONS, AND RELATED TERMS: Maximum Service Temperature

Solidus Temperature

Common Abbreviation:	(None established)
Preferred or SI Unit	°C
Alternate or English Unit	°F

DEFINITION:

The Solidus Temperature is the temperature at which a noneutectic alloy finishes freezing on cooling, or begins to melt on heating. The Solidus is the lower temperature of a material's Melting Range. Unless otherwise stated, the Solidus is reported at standard pressure (1 atm).

EXAMPLE TEST METHOD: ASTM E 1142

MATERIAL/TEST PARAMETERS: Pressure

CONVERSION FACTORS:

to convert	to	multiply by	then add
°C	°F	1.8	32
°F	°C	0.55556	–17.7778
K	°C	1	–273.15
R	°C	0.55556	273.15

Specific Heat Capacity

Common Abbreviation:	c
Preferred or SI Unit	J/kg · K
Alternate or English Unit	cal/g · °C

DEFINITION:

Specific Heat is the Heat Capacity per unit mass, or the quantity of heat required to provide a unit temperature increase to a unit mass of material. The subscript p is used if the determination is at constant pressure. The subscript v is used if the determination is done at constant volume.

EXAMPLE TEST METHOD: ASTM D 2766

MATERIAL/TEST PARAMETERS: Temperature, pressure, volume

CONVERSION FACTORS:

to convert	to	multiply by
J/kg · K	cal/g · °C	0.000239
cal/g · °C	J/kg · K	4,184
Btu/lbm · °F	J/kg · K	4,186.8

SYNONYMS, SYMBOLS, ABBREVIATIONS, AND RELATED TERMS: Specific Heat, Specific Energy Capacity, Specific Entropy, Heat Capacity per Unit Mass, c_p, c_v

EXAMPLE VALUES:	**J/kg · K**	**cal/g · °C**
Carbon and Alloy Steels	450	0.108
Aluminum Alloys	900	0.215
Copper Alloys	385	0.092
Magnesium Alloys	1,050	0.251
Stainless Steels	435	0.104
Titanium Alloys	528	0.126

Thermal Conductivity

Common Abbreviation:	λ
Preferred or SI Unit	W/m · K
Alternate or English Unit	Btu/(h · ft · °F)

DEFINITION:

Thermal Conductivity characterizes the ability of a material to transfer heat; it is the time rate of heat flow, under steady conditions, through unit area, per unit temperature gradient in the direction perpendicular to the area. In other words, Thermal Conductivity (λ) is the constant of proportionality relating the steady-state heat flux (power per unit area, or W/m^2) to the temperature gradient (dT/dx, or K/m) in a material, as expressed by the equation: Power per Unit Area = – λ dT/dx.

EXAMPLE TEST METHODS: ASTM E 1225, ASTM E 1316

MATERIAL/TEST PARAMETERS: Temperature

CONVERSION FACTORS:

to convert	to	multiply by
W/m · K	Btu/(h · ft · °F)	0.5779
Btu/(h · ft · °F)	W/m · K	1.731
(Btu · ft)/(h · ft^2 · °F)	W/m · K	1.731
Btu · in.)/ (h · ft^2 · °F)	W/m · K	0.144
(Btu · in.)/ (s · ft^2 · °F)	W/m · K	519.2
cal/(cm · s · °C)	W/m · K	418.4
cal/(cm · h · °C)	W/m · K	0.116
kcal/(m · h · °C)	W/m · K	1.162
W/m · °C	W/m · K	1

SYNONYMS, SYMBOLS, ABBREVIATIONS, AND RELATED TERMS: k

EXAMPLE VALUES:	**W/m · K**	**Btu/(h · ft · °F)**
Carbon and Alloy Steels	65	37.5
Aluminum Alloys	209	121
Copper Alloys	397	229
Magnesium Alloys	167	96.5
Stainless Steels	16.3	9.4
Titanium Alloys	16	9.2

Thermal Diffusivity

Common Abbreviation:	a
Preferred or SI Unit	m^2/s
Alternate or English Unit	ft^2/s

DEFINITION:

Thermal Diffusivity is defined as the Thermal Conductivity divided by the Specific Heat Capacity and the Density. Thermal Diffusivity is a function of temperature in that Thermal Conductivity, Density, and Specific Heat Capacity are functions of temperature.

EXAMPLE TEST METHOD: ASTM E 1461

MATERIAL/TEST PARAMETERS: Temperature

CONVERSION FACTORS:

to convert	to	multiply by
m^2/s	ft^2/s	10.76426
ft^2/s	m^2/s	0.0929
cm^2/s	m^2/s	0.0001
$in.^2/s$	m^2/s	0.000645

SYNONYMS, SYMBOLS, ABBREVIATIONS, AND RELATED TERMS:
Diffusivity, Thermometric Conductivity

EXAMPLE VALUES:	m^2/s	ft^2/s
Carbon and Alloy Steels	18×10^{-6}	19.4×10^{-5}
Aluminum Alloys	86×10^{v6}	92.5×10^{-5}
Copper Alloys	115×10^{-6}	124×10^{-5}
Magnesium Alloys	91×10^{-6}	98×10^{-5}
Stainless Steels	4.7×10^{-6}	5.06×10^{-5}
Titanium Alloys	6.7×10^{-6}	7.2×10^{-5}

Vapor Pressure

Common Abbreviation:	(None established)
Preferred or SI Unit	Pa
Alternate or English Unit	torr

DEFINITION:

Vapor Pressure is the equilibrium gas pressure of a solid or liquid phase at a given temperature.

EXAMPLE TEST METHOD: ASTM E 1194

MATERIAL/TEST PARAMETERS: Temperature

CONVERSION FACTORS:

to convert	to	multiply by	to convert	to	multiply by
Pa	torr	0.0075	ft H_2O 39.2 °F	Pa	2,988.9
torr	Pa	133.3	in. Hg 60 °F	Pa	3,376.85
mm Hg 0 °C	Pa	133.3	bar	Pa	100,000
mm H_2O 0 °C	Pa	9.807	dyne/cm^2	Pa	0.1
cm H_2O 0 °C	Pa	98.07	atm (normal = 760 torr)	Pa	101,325
in. H_2O 32 °F	Pa	249.09	atm (tech = 1 kgf/cm^2)	Pa	98,066.5

Index

A

B

C

D

E

F

G

H

I

J - L

M

S

T

U

V

W- Z